ROUTLEDGE LIBRARY EDITIONS:
WOMEN AND BUSINESS

Volume 4

PHOSSY JAW AND THE FRENCH MATCH WORKERS

PHOSSY JAW AND THE FRENCH MATCH WORKERS

Occupational Health and Women in the Third Republic

BONNIE GORDON

Routledge
Taylor & Francis Group

LONDON AND NEW YORK

First published in 1989 by Garland Publishing, Inc.

This edition first published in 2017
by Routledge
2 Park Square, Milton Park, Abingdon, Oxon OX14 4RN

and by Routledge
711 Third Avenue, New York, NY 10017

Routledge is an imprint of the Taylor & Francis Group, an informa business

British Library Cataloguing in Publication Data
A catalogue record for this book is available from the British Library

ISBN: 978-1-138-23710-0 (Set)
ISBN: 978-1-315-27106-4 (Set) (ebk)
ISBN: 978-1-138-24516-7 (Volume 4) (hbk)
ISBN: 978-1-138-28083-0 (Volume 4) (pbk)
ISBN: 978-1-315-27183-5 (Volume 4) (ebk)

Publisher's Note
The publisher has gone to great lengths to ensure the quality of this reprint but
points out that some imperfections in the original copies may be apparent.

Disclaimer
The publisher has made every effort to trace copyright holders and would welcome
correspondence from those they have been unable to trace.

Phossy Jaw
and the French Match Workers

Occupational Health and Women
in the Third Republic

Bonnie Gordon

GARLAND PUBLISHING, INC.
New York   London
1989

Copyright © 1989 by Bonnie Gordon.

Library of Congress Cataloging-in-Publication Data

Gordon, Bonnie.
Phossy jaw and the French match workers : occupational health and
women under the Third Republic / Bonnie Gordon.
p. cm. — (Garland studies in historical demography)
Thesis (Ph. D.)—University of Wisconsin Madison, 1985.
Includes bibliographical references.
ISBN 0-8240-3363-9 (alk. paper)
1. Match industry workers—Diseases—France—History. 2. Match
industry workers—Health and hygiene—France—History. 3. Jaws—
Necrosis—Prevention—Government policy—France—History. 4.
Industrial hygiene—Government policy—France—History. 5. Trade-
unions—Match industry workers—France—History. 6. Women—
Employment—France—History.
I. Title. II. Series.
HD7269.M282F846 1989
363.11'96625'0944—dc20   89-35154

Printed on acid-free, 250-year-life paper

Manufactured in the United States of America

Phossy Jaw
and the French Match Workers

Occupational Health and Women
in the Third Republic

TABLE OF CONTENTS

ADDENDA

MAPS, CHARTS AND GRAPHS

In the following charts and graphs, all figures on match workers reflect
my sample of the communes of Pantin-Aubervilliers unless otherwise
specified.

CHAPTER I: INTRODUCTION: OCCUPATIONAL HEALTH
THE STATE IN NINETEENTH CENTURY FRANCE

The 1898 suppression of white phosphorous in the French match industry was a victory of organized labor. It was a significant step forward because it was one of the first bans on any industrial poison. Thus, the story sheds light on how organized workers can break new ground in health-related workplace reform. It was significant because France was the second country to ban white phosphorous although all match manufacturing countries were aware of the problem.[1] After France banned white phosphorous and substituted a non-toxic formula, most of the other match manufacturing countries followed suit. Thus, the International Accord of Berne, signed in 1906 by France, the German Empire, Denmark, Italy, Luxemburg, Holland, and Switzerland, was also a victory due to French organized labor.[2] The suppression of white phosphorous was also especially significant because the majority of the workers in the match industry and the majority of the members of the Match Workers' Federation were women. At a time when most French workers did not have the power to effect changes in the health and safety conditions of their work, the match workers succeeded. At a time when most French working women were not unionized and did not pursue effective action on occupational health problems, French women in the match industry succeeded. Thus, the story sheds light on what sort of working women were willing to take concerted action as well as how those women's participation in labor union action was different from the participation of men.

This study is also a correction to the two prevailing misinterpretations of how France arrived at the ban and the substitution of a non-toxic match. The first misinterpretation gives credit to the foresight and humanity of the Republican State. No less a knowledgeable source than Léon and Maurice Bonneff, authors of several works on dangerous trades at the turn of the century, wrote in 1913:

> "One need not hesitate to praise the Republican State for making its first consideration, as soon as it had decided to assume direct control over match manufacturing, to assure the healthfulness of its factories and the harmlessness of its products. Well-lit and well-ventilated workshops were built. Research proceeded non-stop to replace white phosphorous..."[3]

On the contrary, this study shows that the Republican State procrastinated despite its monopoly of the industry and the recommendations of its most illustrious doctors until the French match workers forced its hand.

The second misinterpretation gives the credit to the two chemists who invented the non-toxic formula. On the contrary, this study shows that inventors do not work in a historical vacuum. Serious experimentation was not even encouraged until after the French match workers had exerted so much pressure that a ban, with or without a non-toxic substitute, had to be imposed.

To understand the particular problem of white phosphorous necrosis in the French match industry, it is necessary to begin with the general problem of working class health on and off the job in the nineteenth century. Workers were able to secure improvements only in a few special situations where they could touch an economic pressure point or a social pressure point. Miners, for example, secured protective legislation because of their concerted action in an economically strategic industry. Women and children were the recipients of protective legislation because of their social functions, not their labor functions. Other piecemeal improvements came about in isolated industries because a social need, the protection of the public health, was judged to override the economic need to exploit private property. However, the story of the match workers' struggle was played out at a time of a growing social awareness of some responsibility to provide for those workers who were injured or killed on the job. From 1880 to 1898 Parliament considered bills on workers' compensation, finally passing a law on April 9, 1898. Certainly, the fact that some workers were singled out for special protection for economic or social reasons must be seen in the political context of a Republican legislature which was trying to recognize society's responsibility for public health.

Throughout the nineteenth century, reformers warned that industry was ruining the health of the French working class. In Louis-René Villermé's 1838 investigation of cotton, wool, and silk workers, he introduced all of the occupational health themes of France's first industrial century. He found, in essence, that the traditional agricultural life was good and that the manufacturing life was bad.

"One must see the multitude of lean, pale, ragged children who go to work bare-footed in the rain and the mud..."[4]

After this description of the children of Mulhouse, he compared

2

"the florid complexions, plumpness, spirit, and all the signs of excellent health that one notices in children of the same age every time one leaves a manufacturing area to enter an agricultural canton."[5]

Whether working in a factory, a workshop, or as a pieceworker at home, one worked long hours that ensured exhaustion and precluded any time for cooking or eating complete meals or cleaning house; Rouen and Reims weavers performed twelve to seventeen hours a day of labor. Women were paid so little that many, like the spinners of Reims, resorted to part-time prostitution, known as "doing the fifth quarter of the day". Children who were set to work as young as age four and a half grew up stunted. In Amiens, Villermé found that 66% percent of working class military recruits were rejected for poor health or weak constitutions but only 44% of the sons of "people of means" were rejected for the same reasons. Many children died young due to inadequate nutrition and care. Among infants at the *Hotel — Dieu* of Reims, 64% died in their first year; among those raised at home, only 50% died. Lille cotton workers had phthisia, which often meant tuberculosis: a disease of poverty, malnutrition, and lack of sunlight. Phthisia was a general term used to describe any disease of the lungs: general congestion from breathing cotton dust, fibrosis, black lung, or cancer. Weavers in Lyon had painful chests and abdomens because the balance wheel of their looms tapped constantly in one spot. Silkworkers complained that their work was dirty, foul-smelling, and required plunging the fingers into nearly boiling water. Various occupations gave rise to skin diseases and rheumatism. In all occupations, the workers were unprotected against accidents.[6]

In 1838 Villermé knew the problems and the solutions. To prevent accidents put safety guards around the dangerous parts of machinery and make it law so that all manufacturers comply. To improve the general health, raise the starting age, lower the number of hours in the workday, and improve living conditions.

Fifty years later, conditions had hardly changed. "The worker always labors with death hanging over his head," declared the miners of the Tarn in 1883. A miner could expect to live an average of 54 years but a glassblower could expect only 37.[7] As late as 1900 only 30% of all working class adults lived to age 65 according to the *Parti socialiste*.[8]

Therefore, the match workers' struggle against white phosphorous necrosis in the 1890s must be seen as only one event in one small industry of 2,000-20,000 workers. It took place at a time when lead and mercury poisoned many more workers in a variety of industries, and when crippling and killing accidents occurred to uncounted thousands.[9] Among the bourgeoisie and the working class, there was rarely a clearly understood dividing line between the ill health which was induced by work and the ill health which was induced by inadequate housing, cleanliness, and nutrition. On the part of the bourgeoisie, it must be added that there was the general assumption that workers' ill health was induced by the workers' own immorality and character defects. Workers who lost their lives or limbs to machinery were called careless. Women workers were accused of looking at themselves in pocket mirrors instead of minding their machinery. When fires broke out, workers were accused of violating work rules which they had no time to observe.[10] To the class in power, occupational health was a non-problem. In isolated cases, it was considered to be a question of bad moral character.

From the workers' point of view, other struggles came first. Of the 1,997 strikes from 1871-1890 which were analyzed by Michelle Perrot, only 1.6% concerned health, safety, compensation, or retirement.[11] Resolutions from union congresses indicated that workers believed that the best way to improve their health was to work shorter hours and earn more money; therefore, they would be less tired and have more choice about food, clothes, and lodging. Hence, the workers' movement and their socialist allies took up the cry for the eight-hour day.[12]

The first few exceptions to this bleak picture occurred in those trades with one of the following three structural conditions. First, the workers might be in a strong enough position to exert economic pressure on their employers in favor of compensation after accidents. In a second category, women and children were singled out as weak members of society who were unable to protect themselves; therefore, they were considered suitable recipients of legal protection. In a third category, those industries which came to the attention of the public health authorities, called the *conseils de salubrité*,[13] might be made to take measures in the interest of the general public. These measures were unlikely to be enforced if they were too costly or required any significant reorganization of labor.

In the first category were miners. The organization of labor in mines favored solidarity; the frequency of accidents and the necessity of collective action to aid the victims increased solidarity. The

4

isolated locations of most mines made the importing of strikebreakers difficult. Therefore, coal miners were among the first to form a national federation, in 1883. They elected the Secretary of their Federation, Émile Basly, a forceful personality and tireless advocate, to the Chamber of Deputies in 1885. Furthermore, the importance to the national economy of coal and iron mining in particular placed those miners in a strong position vis-a-vis their bosses. Consequently, they secured accident compensation funds and pensions earlier than other French workers, in 1894. In the same year, miners secured recognition of their special status in a safety law which mandated the participation of delegates from the Miners' Federation.

A second wedge driven into nineteenth-century liberal capitalism was protection for women and children. This took the form of limitations on the hours of labor of women and children and measures to preserve the lives of children. Although it was contrary to the prevailing theory of economic liberalism to limit an adult male's freedom to work himself to death, even the staunchest liberals agreed that children did not enter freely into agreements as the equals of their employers.[14] The legal status of women was more complicated; hence legislation in their favor was slower to evolve. On the one hand, they might at any time be carrying a potential citizen who had not agreed to be poisoned or weakened in the womb. The presence of a mother at any workplace outside of her home was considered detrimental to her children's well-being. Added to the concerns about the welfare of children was always the suspicion that any woman who was not under her parents' or her husband's roof was engaging in illicit sex.[15] Bourgeois reformers' criticisms of the industrial employment of women rose to a crescendo in the last years of the century, resulting in two pieces of hours legislation: the laws of 1874 and 1892.

However, by the last quarter of the century, the reality of French women's presence in the workforce became accepted in the writings of reformers. In Paul Leroy-Beaulieu's *Le Travail des femmes au dix-neuvième siècle*, he argued that it was impossible to exclude women from the workforce because the organizing principle of society was the individual, not the tribe. By this he meant that if a woman had neither a father nor a husband to support her, or if her father or husband could not support her, she could not legally oblige her brother, uncle, or cousin to help. She had to work. Leroy-Beaulieu went on to cite studies which proved that sexual freedom and illegitimate births were less frequent among women who worked

5

in large industrial plants rather than in artisanal shops, agriculture, or piecework performed at home.[16] At the turn of the century, studies appeared which were no longer moral treatises but factual accounts of where women fit into the workforce.[17]

When protection of health was discussed, the emphasis was on the health of those children whose mothers worked. There was concern over who was watching and feeding the toddlers while the woman and her older children were away from home. Many were the accounts of dishonest wet nurses who neglected the babies of the poor.[18] There were also mothers who doped or tied their children to keep them out of trouble while the mothers were at work.

To the extent that the health of the working woman was discussed at all, it was her "moral health," a euphemism for her sexuality, rather than her freedom from disease. Throughout the century, bourgeois ideologists, including doctors, hardly understood to what extent illnesses were caused by bad air, bad water, contagion, or internal disorders. While they noted that working class women led lives of overwork relieved only by drink and sex, reformers could not stop ascribing the general weakness and sorry looks of the working women to their wanton ways rather than to the factory life.

Society's protection was timidly extended by the laws of 1841 and 1851, which gave working class children their first eight years and ten years, respectively, free to grow into future citizens before harnessing them to machinery for fourteen or more hours a day. It was extended under the Third Republic by the law of May 19, 1874, which limited the hours of children to six a day and the hours of young women to twelve a day. This law was applicable to large workshops where most women and children were not employed.[19] Even so, this effort was the fruit of considerable bourgeois dismay at the discovery that women and children were working in industry at all. Similarly, the first piece of protective legislation in the match industry applied to children. By a decree of September 22, 1879, the Minister of Commerce forbade the employment of children under the age of twelve in match factories because of the danger of fires.[20]

Inadequately enforced by very few work inspectors, the law of 1874 remained the only official protection for women and children until 1892. In the following year for the first time, republicans, radicals, and socialists held the majority in the Chamber of Deputies.[21] Radicals and socialists were anxious to respond to class struggle. For this reason, they passed another hours law on November 2, 1892 and

6

a public assistance law for pregnant women on July 15, 1893. The legal limit to the workday was reduced to eleven hours for adult women and ten hours for children. The age of the youngest worker was raised to twelve with an elementary school certificate, age thirteen without one. This law repeated the 1874 ban on night work and mandated one weekly day of rest.

Although the few improvements which were made were beneficial only to mineworkers, women, and children, there were other, piecemeal efforts throughout the century but only where hygiene measures did not interfere with the organization of the labor process itself and were not too costly. This was the case in those industries which were recognized as public nuisances by local health councils.

In the literature on unhealthy workplaces, the words "aeration" and "ventilation" occurred incessantly. This was not only because bad smells were the most obvious evidence to the public of unhealthy industries, but also because nineteenth century medicine was still influenced by the theory of "effluvia": allegedly disease-causing substances which were carried in foul-smelling air.[22] When the coalition which was elected to Parliament in 1893 passed another piece of progressive legislation on June 12, 1893 (See Addenda), it contained surprisingly specific and generous clauses on cubic meters of air per worker; the removal of gases and dusts; the frequency of cleaning floors, walls, and ceilings; the separation of lunchrooms from worksites; and the availability of washbasins and clothes-changing facilities. This law, however, did not call for any changes in the organization of labor as the hours limitation law did. It only codified what many large factories were already providing and what some owners of large factories desired in order to equalize their expenses with the expenses of their competitors. However, as a law to improve the healthfulness of factories, it failed. Because inspection was rare and ineffective, small manufacturers had no incentive to spend money on fans and ventilation or to build separate locker rooms and lunch rooms. Although this law did not interfere with the organization of production, it cost too much to be universally applied.

What public health councils were able to impose were rules on the locations of dangerous or unhealthy industries. Often the police prefect would take the health council's advice and refuse a permit to a workshop or factory unless the owner moved it to the outskirts of town. Since land was usually cheaper there, it was a solution which caused little financial hardship to manufacturers.

7

The match industry in the 1890s was a focal point for occupational health improvements because it united the first two conditions: a strong workers' organization and the presence of women and children. The workers themselves brought about the third condition: the attention of *conseils de salubrité* to a solution which became cheaper than the problem. First, the match workers formed the National Federation of Match Workers in 1892, uniting all match factories. Because of the political conjuncture of the 1890s and the match workers' special status as government workers, they were in an excellent position to bargain with the Ministry of Finances and Parliament. Second, the public and elected officials were further impressed by the spectacle of women and children in an industry which mutilated the face and was said to interfere with child-bearing. Third, the change which the match workers wanted was not expensive but it did require some research funds. The match workers, with the aid of certain medical doctors, made it worth the Administration's while to spend a little money on research rather than a lot of money on sanitation and sickpay.

However, during the 1890s Parliament was also considering a workers' compensation law. Lest this seem to contradict the bourgeois indifference described above, it must be explained that the workers' compensation law of April 9, 1898 was not a concession to working class health. It did not mandate improvements in health and safety conditions but merely provided a percentage of the injured worker's wages from the fourth day of disability.[23] Most employers soon subscribed to the *Caisse Nationale d'Assurances* and passed the premiums on to consumers.

Indeed, by the 1890s, major industrialists like Henri Darcy, who was the president of the Central Committee of Coal Mines and the president of the ironworks of Chatillon, Commentry, and Neuves-Maisons, was in favor of some form of workers' compensation law. This was because the Civil Code named the owner responsible for any accident which occurred on his premises whether it was caused by negligence or error on the part of the owner, a manager, or a worker. Mechanization itself, Darcy argued, was responsible for many accidents which had not been foreseen when the Civil Code was written during the First Empire. Industrialists, therefore, wanted a law which would spare them the lawsuits which were sometimes filed by a badly disabled worker or the dependents of a deceased worker.[24]

The workers' compensation law was not a response to a demand by labor. Where trade union congresses passed accident compensation resolutions, they figured at the end of the list. Their language mirrored that of whichever bill was currently before Parliament. In addition to high levels of compensation, workers wanted their bosses to be subject to fines. The left press did not make a campaign of the accident compensation bills which moved through Parliament from 1880-1898. Instead, the press carried only short notices on work accidents under *faits divers*, usually without any comment on causes or remedies. The literature on work and health, whether memoirs of workers or studies on the left,[25] indicated a fatalistic attitude towards occupational health and safety: there was little one could do on the job to avoid injury or illness. Workers' participation in the efforts to pass the workers' compensation law did not take the form of a militant movement comparable to the miners' efforts to secure their legislation in 1894.

Nor was the law of 1898 an effort to protect women and children. It was not directed to small workshops and sweated labor, which is where women and children were most numerous. Rather, it was applicable to large, motorized or steam-powered worksites. Neither did it address occupational diseases, which were a greater threat to women than were industrial accidents.

However, the French working class in general and the match workers in particular benefited from the legal acceptance of the concept of occupational risk, *le risque professionnel*.[26] This was part of every workers' compensation bill from 1885 to the final version in 1898. This was a blow struck against economic liberalism. The 1898 law bore fruit in the twentieth century as its provisions were extended to broader categories of workers than factory laborers. By 1919 it was extended to cover some occupational diseases. The theory of occupational risk, therefore, added a cost consideration to capitalists' and factory managers' budgets. They knew by the 1890s that they would soon have to include health insurance costs or improvements in plant and equipment in their operating expenses.

The match workers also benefited from the acceptance by jurists of the principle of occupational risk. Although the State's legal obligation to compensate the victims of occupational health and safety hazards was vague, it was becoming obvious that the principle would be voted into law during those same years when the match workers were posing their demands: 1888-1898. The State administrators of the match industry acted when they did partly because of fear that compensation for white phosphorous necrosis would cost more in future years if they did not stop the problem.

ADDENDA

LOI DU 12 JUIN 1893

concernant l'hygiène et la sécurité des travailleurs
dans les établissements industriels

Article 1. Sont soumis aux dispositions de la présente loi: les manufactures, fabriques, usines, chantiers, ateliers de tous genres et leurs dépendances.

Sont seuls exceptés: les établissements où ne sont employés que les membres de la famille, sous l'autorité soit du père, soit de la mère, soit du tuteur.

Néanmoins, si le travail s'y fait à l'aide de chaudière à vapeur ou de moteur mécanique, ou si l'industrie exercée est classée au nombre des établissements dangereux ou insalubres, l'inspecteur aura le droit de prescrire les mesures de sécurité et de salubrité à prendre, conformément aux dispositions de la présente loi.

Article 2. Les établissements visés à l'article 1er doivent être tenus dans un état constant de propreté et présenter les conditions d'hygiène et de salubrité nécessaires à la santé du personnel.

Article 3. Des règlements d'administration publique, rendus après avis du Comité consultatif des arts et manufactures, détermineront:

Dans les trois mois de la promulgation de la présente loi, les mesures générales de protection et de salubrité applicables à tous les établissements assujettis, notamment en ce qui concerne l'éclairage, l'aération ou la ventilation, les eaux potables, les fosses d'aisance, l'évacuation des poussières et vapeurs, les précautions à prendre contre les incendies, etc.

Au fur et à mesure des nécessités constatées, les prescriptions particulières relatives, soit à certaines industries, soit à certains modes de travail. Le Comité consultatif d'hygiène publique de France sera appelé à donner son avis en ce qui concerne les règlements généraux prévus au paragraphe 3 du présent article.

DUCRET DU 10 MARS 1894 CONCERNANT L'HYGIENE ET LA SECURITE DES TRAVAILLEURS DANS LES ETABLISSEMENTS INDUSTRIELS

(Application de l'article 3 de la loi du 12 juin 1893)

Article 1. Les emplacements affectés au travail dans les manufactures, fabriques, usines, chantiers, ateliers de tous genres et leurs dépendances seront tenus en état constant de propreté. Le sol sera nettoyé à fond au moins une fois par jour avant l'ouverture ou après la clôture du travail. Ce nettoyage sera fait soit par un lavage, soit à l'aide de brosses ou de linges humides, si les conditions de l'industrie ou la nature du revêtement du sol s'opposent au lavage. Les murs et les plafonds seront l'objet de fréquents nettoyages; les enduits seront refaits toutes les fois qu'il sera nécessaire.

Article 3. Les travaux dans les puits, conduites de gaz, canaux de fumée, fosses d'aisances, cuves ou appareils quelconques pouvant contenir des gaz délétères, ne seront entrepris qu'après que l'atmosphère aura été assainie par une ventilation efficace. Les ouvriers appelés à travailler dans ces conditions seront attachés par une ceinture de sûreté.

Article 5. Les locaux fermés affectés au travail ne seront jamais encombrés: le cube d'air par ouvrier ne pourra être inférieur à six mètres cubes.

Ils seront largement aérés. Ces locaux, leurs dépendances et notamment les passages et escaliers, seront convenablement éclairés.

Article 6. Les poussières ainsi que les gaz incommodes, insalubres ou toxiques, seront évacués directement au dehors de l'atelier au fur et à mesure de leur production.

Pour les buées, vapeurs, gaz, poussières légères, il sera installé des hottes avec cheminées d'appel ou tout autre appareil d'élimination efficace.

Pour les poussières déterminées par les meules, les batteurs, les broyeurs, et tous autres appareils mécaniques, il sera installé, autour des appareils, des tambours en communication avec une ventilation aspirante énergique.

Pour les gaz lourds, tels que les vapeurs de mercure, de sulfure de carbone, la ventilation aura lieu *par descensum*: les tables ou appareils de travail seront mis en communication directe avec le ventilateur.

La pulvérisation des matières irritantes ou toxiques ou autres opérations, telles que le tamisage ou l'embarillage de ces matières, se feront mécaniquement, en appareils clos.

L'air des ateliers sera renouvelé de façon à rester dans l'état de pureté nécessaire à la santé des ouvriers.

Article 7. Pour les industries désignées par arrêté ministeriel, après avis du Comité consultatif des arts et manufactures, les vapeurs, les gaz insalubres et incommodes et les poussières seront condensés ou détruits.

Article 8. Les ouvriers ne devront pas prendre leurs repas dans les ateliers ni dans aucun local affecté au travail.

Les patrons mettront à la disposition de leur personnel les moyens d'assurer la propreté individuelle: vestiaires avec lavabos, ainsi que l'eau de bonne qualité pour la boisson.

Article 9. Pendant les interruptions de travail pour le repas, les ateliers seront évacués et l'air sera entièrement renouvelé.

ENDNOTES CHAPTER I

1. Denmark banned white phosphorous matches in 1874 because of the accidental and deliberate poisonings of consumers, the use of matches in provoking abortions, the frequency of fires, and the danger to the workers. Switzerland imposed a ban in January 1881. The Swiss public resorted to contraband white phosphorous matches. The ban was rescinded in June 1882.

2. In 1908 Great Britain signed; in 1909 Spain joined.

3. Léon and Maurice Bonneff, "Chez les 'Bouts de Bois' d'Aubervilliers," *l'Humanité*, March 5, 1913.

4. Dr. Louis-René Villermé, *Tableau de l'état physique et moral des ouvriers employés dans les manufactures de coton, de laine, et de soie*, Paris: Union générale d'éditions, 1971, 229.

5. idem

6. ibid., 62-271.

7. Strikers to the Prefect of the Tarn, March 8, 1883, Archives départementales Tarn IV M268 cited in Rolande Trempé, *Les Mineurs de Carmaux 1848-1914*, tome I, Paris: Les éditions ouvrières, 1971, 324 and 583.

8. Peter Stearns, *Revolutionary Syndicalism and French Labor: A Cause without Rebels*, New Brunswick: Rutgers University Press, 1971, 263.

9. Accidents were uncounted because, in the absence of a workers' compensation plan, there was no one who was responsible for keeping figures.

10. Hector Malot, *En famille*, Paris: Editions G.P., 1960, 209. In an 1893 incident at the Aubervilliers match factory, a woman dropped matches on the floor. Rather than pick them up, according to the work rules, thus slowing down her productivity and reducing her wages, she stepped on the matches. When they burst into flames, her skirts caught on fire. She then left her workshop without authorization, in further violation of the rules, to run into the courtyard. This fanned the flames. Three days later she died of burns. Management at first refused to pay compensation to her orphan on the grounds that she had violated the rules. Archives de la Préfecture de Police, Paris, 1.408 bis.

11. Michelle Perrot, *Les Ouvriers en Grève*, Paris and The Hague: Mouton, 1974, volume I, 259.

12. In February 1892 *Le Socialiste* ran a series of articles entitled "Le Droit à l'Hygiène" or "The Right to Health". The conclusion to

each article was that the eight-hour day was the way to improve health.

13. In each *département* the police had the right to call on an advisory board of doctors as public health authorities: the *conseil de salubrité* or *conseil d'hygiène*. The police chose the doctors, decided how often they would meet, and selected the questions they would address. The police prefect then forwarded their recommendations to the Ministry of the Interior, which had the sole authority to authorize action. Inspection was rare, even after the passage of the 1874 law which restricted the labor of women and children, and the 1892 law on the general health and safety of manufactures.

14. Paul Leroy-Beaulieu, *L'Etat moderne et ses fonctions*, Paris: Guillaumin, 1891, 339. He justified limiting children's hours in order to protect them against the exploitation of greedy parents.

15. "A woman's virtue is like down feathers; if she works in a workshop, first it is soiled, then it disappears. Do not expose her to contacts". Denis Poulot, *Le Sublime*, Paris: Maspéro, 1980, 261.

16. Paul Leroy-Beaulieu, *Le Travail des femmes au dix-neuvième siècle*, Paris: Charpentier, 1873.

17. Archives Nationales F22 333. Rapport présenté à Monsieur le Président de la République par MM. les membres de la Commission Supérieure du Travail des Enfants et des Filles Mineures Employées dans l'Industrie, 1884; Jeanne Chauvin, *Etude historique sur les professions accessibles aux femmes*, Paris: Giard and Brière, 1892; Vallier, *Le Travail des femmes dans l'industrie francaise*, 1899; Benzacar, "L'Ouvrière au XXe. siècle", *Questions pratiques*, 1902; Dr. Kaethe Schirmacher, *Le travail des femmes en France*, Paris: Rousseau, 1902; Froment, *Ouvrières parisiennes*, Lille: Action populaire, 1903; René Gonnard, *La femme dans l'industrie*, Paris: Colin, 1906; Fénélon Gibon, *Employées et ouvrières*, Lyon: E. Vitte, 1906; Caroline Milhaud, *L'ouvrière en France. Sa Condition présente*, Paris: Alcan, 1907. Of particular interest are the specialized studies of the Office du Travail on the sweated trades where women were most numerous: Ministre du Commerce, de l'Industrie, et des Colonies. Office du Travail, *La Petite industrie. Salaire et durée du travail*, 1893; *Le vêtement à Paris*, 1896; *L'industrie du chiffon à Paris*, 1903; *Enquête sur le travail à domicile dans l'industrie*

de la fleur artificielle, 1913; *Enquête sur le travail à domicile dans l'industrie de la lingerie*, 1917.

18. This is a recurring theme in Emile Zola, *Fécondité*, Paris: Charpentier, 1915. Single working mothers paid a fixed fee for the care of their children until first communion, knowing that they would never see the children alive again.

19. Women and children were also barred from working from nine p.m. to five a.m. There was little enforcement.

20. Direction Générale des Manufactures de l'Etat, *Lois, Ordonnances, Décrets, Décisions, et Arrêtés Ministeriels*, Paris: Imprimerie Nationale, 1894, 175.

21. Fifty socialist deputies and one hundred and forty radicals were elected in 1893.

22. Robert Koch proved that certain diseases were caused by microorganisms in 1882.

23. Figures on industrial accidents before 1900 were kept only for the merchant marine, for railroads, and for mines. From 1900-1911 the figures on workers killed or injured rose. In 1900, 1,735 were killed and 231,241 were injured. In 1911, 2,002 were killed and 472,394 were injured. Ministère du Commerce, *Annuaire statistique de la France*, Paris: Imprimerie Nationale, 1912, Volume 32, 130 (hereafter called *Annuaire statistique*).

24. Henri Darcy, "Etat actuel de la question des accidents du travail", *Congrès des accidents du travail*, Paris: 1898.

25. Jeannette Laot, *Stratégie pour les femmes*, Paris: Stock, 1977; Lucie Baud, "Mémoires", *Le Mouvement socialiste*, June 1908, 418-425 reproduced in *Le Mouvement social*, 1978 number 105, 139-146; Caroline Milhaud, *L'ouvrière en France. Sa condition présente. Les réformes necéssaires*, Paris: Alcan, 1907; Fernand and Maurice Pelloutier, *La Vie Ouvrière en France*, Paris: Schleicher, 1900.

26. "Risk which is inherent to the occupation, the industry, outside of any fault of the employer or the worker", Paul Nourrisson, *Le risque professionnel et les accidents du travail*, Paris: Larose and Forcel, 1891, 4.

CHAPTER II: THE HISTORY OF THE MATCH INDUSTRY, OF PHOSSY-JAW, AND OF EFFORTS TO COMBAT IT UNTIL 1890

Matches were invented in the 1830s.[1] According to an encyclopedia used in French schools in the 1880s (see illustration), the process was as follows. Poplar, aspen or pine logs were cut across the trunk into slices (still called *camemberts* today). Each of these was cut into several hundred small sticks which were then arranged by hand in four-sided topless and bottomless presses. Each press was lowered into a potassium or sulphur bath and then into a phosphorous bath so that only the head of each stick absorbed the chemicals: this step was called the dipping or the chemical bath. The sticks were then moved elsewhere to dry, removed from the presses, and picked over by hand to remove imperfect ones. The rest were packaged. Sometimes fancy labels were added.[2]

Obviously, the workers who cut the wood or set up the presses were not exposed to white phosphorous. The dippers, however, stood over basins of warm phosphorous, breathing fumes and splashing phosphorous over their hands and clothes for twelve or more hours every day. The workers in the removal and the packaging sections were also exposed, although less so, because they handled the mixture when it was cool and no longer dripping. Throughout the process, ventilation was the only means of diminishing exposure. Where the workshops were separated, there was also less exposure of all the workers outside of the dipping area. However, where partitions between work areas were not available, and where ventilation was poor, all of the workers were exposed all of the time.

At first a curiosity, matches quickly became a common article of consumption. For example, the *Almanach de Commerce* for the *département* of the Seine listed only three manufacturers in 1836. Their quaint descriptions of their product indicated the novelty. One explained:

> "one makes a fire by passing the match quickly across a sandy paper..."

Another advertised:

> "striking matches called Lucifers and flameless matches for smokers."

The third had:

"wax pyrogenous matches, electric, magic, and infernal matches."[3]

By 1845 there were nine manufacturers listed. Ten years later match manufacturing had become a well-established industry with twenty-two factories listed in the Paris almanach alone.[4] Lyon and the Rhône Valley were also a match producing area. According to an inquiry done by the Ministry of Finances in 1852, there were 150 factories in all of France. By 1870 there may have been as many as 600 according to one source, over a thousand according to another.[5]

The Ministry of Agriculture, Commerce, and Public Works was responsible for the health hazards caused by industry. The Ministry became aware of the dangers in this growing industry through reports from police prefects who were prompted by their departmental health councils. As early as the 1840s, prefects urged the classification of match factories as first or second class health hazards: *établissements insalubres*. However, this was less a response to workers' health needs than a response to complaints of fires and explosions from the neighbors of match factories.

Only ten years after the beginning of production, the first medical reports appeared on necrosis. Any toothache might indicate the beginning of the degenerative and disfiguring disease. Eventually one or more teeth fell out or became loose enough to extract with the fingers. The gums and jaws swelled and hardened, which made chewing so painful that some patients suffered as much from malnutrition and weakness as from necrosis. The patient also gave off a strong offensive odor of phosphorous and often lost the ability to articulate words clearly. Next, pieces of the jawbone broke off and were washed out with pus. The patient might linger in this state for months or years. In the final stage, the jaw could be lifted out by hand. The patient usually survived but with a deformed and scarred face. Occasionally, the bone-rotting extended to the other bones of the face.[6]

There was a second disease which was believed by doctors to be common to all match workers: phosphorism. Its symptoms were yellow skin, dark circles around the eyes, phosphorescent breath, a characteristic odor to the breath, demineralization, easily fractured bones, infertility, miscarriages and stillbirths.[7]

L'HISTOIRE D'UNE

Dans les temps primitifs, comme aujourd'hui encore chez les sauvages, le seul moyen de se procurer du feu était de frotter rapidement l'un contre l'autre deux morceaux de bois très secs, ou de faire jaillir l'étincelle de deux cailloux.

Les Romains ont connu l'allumette soufrée. Au moyen-Âge, chaque bourgeois portait sur lui dans un petit sac son briquet en forme de B. Ce briquet, source de feu, figure dans le blason des ducs de Bourgogne, et c'est de Briquets que se compose le fameux collier de la Toison d'Or.

ALLUMETTE

SÉRIE ENCYCLOPÉDIQUE GLUCQ
des Leçons de Choses Illustrées

Ouvrage adopté par la VILLE de PARIS
comme Récompense dans ses Écoles

Il n'y a pas encore bien longtemps, on ne se servait que de ces briquets primitifs dont l'Étincelle allumait un morceau d'amadou. Les vieux soldats de Napoléon n'ont eu que des FUSILS A PIERRE pour conquérir l'Europe.

Le SOUFRE provient des volcans de Sicile et a été connu de toute antiquité. C'est vers 1774 que Scheële, illustre chimiste suèdois, découvrit le PHOSPHORE en calcinant des os. Pur, le phosphore s'enflamme à l'air libre ! mais mélangé en pâte, il ne s'enflamme plus que par le frottement.

Illustration: The Story of the Match

Telle a été l'idée première de l'allumette chimique. Le phosphore s'enflammant par le frottement, allume le soufre : le soufre allume le bois : et l'allumette s'enflamme à son tour et communique le feu où l'on veut...

La fabrication des allumettes consomme des forêts entières! Les bûcherons choisissent les arbres les moins noueux, les moins durs et les plus combustibles, tels que le peuplier, le tremble et le sapin. Une fois abattus, ces arbres sont expédiés aux fabriques.

Les arbres sont coupés d'abord en rondelles de 6 CENTIMÈTRES d'épaisseur. Puis, ces rondelles sont placées sur une machine munie de 25 lancettes qui, à chaque coup, abattent 300 petits bouts de bois ou allumettes.

Une machine ingénieuse couche tous ces petits bouts de bois par rangs étagés dans une caisse sans fond, où ils sont solidement maintenus et séparés en tranches parallèles par de petites traverses ... une PRESSE. Chaque PRESSE contient 5,500 allumettes.

Dépôt exclusif chez M. A. CAPENDU, 1, Place de l'Hôtel-de-Ville, Paris.

Illustration: The Story of the Match (b)

Cette **PRESSE** est alors plongée dans un bain de soufre liquide, puis passée sur une plaque recouverte d'une épaisse couche de pâte chimique phosphorée. Les 5,500 allumettes se trouvent ainsi soufrées puis phosphorées d'un seul coup.

On laisse sécher les allumettes : puis, lorsqu'elles sont sèches, on les empile sur une machine spéciale qui en laisse tomber, à chaque tour de roue, juste un paquet de cent, qu'un piston vient pousser et faire tomber dans une petite boîte. L'allumette est finie.

On fabrique également des **ALLUMETTES-BOUGIES**. Dans ce cas, les petites bûchettes de bois sont remplacées par des bouts de tresse de coton sans fin, que l'on fait passer dans un bain de cire fondue, et qu'une machine spéciale coupe en bouts de 5 centimètres.

Il faut se rappeler que les allumettes au phosphore ordinaire et au soufre constituent un grave danger : d'abord elles sont un poison violent; puis, à cause des enfants, elles sont l'occasion de trop nombreux incendies dont on entend parler tous les jours.

Paris, — Auteur-Éditeur de la série encyclopédique des Leçons de Choses Illustrées.

Illustration: The Story of the Match (c)

Un industriel français a eu l'idée de fabriquer
des allumettes avec du phosphore **AMORPHE**.
Ainsi fabriquée, l'allumette a perdu ses propriétés
vénéneuses et les enfants peuvent impunément la
porter à la bouche. De plus, elle ne peut plus
s'enflammer que si on la frotte sur un **ENDUIT
SPÉCIAL**. Partout ailleurs, elle ne s'enflamme
pas.

Cette découverte merveilleuse a été bien vite
appliquée, et c'est ce qu'on nomme aujourd'hui les
ALLUMETTES SUÉDOISES, au phosphore
amorphe. Ce sont les seules que l'on emploie dans
les administrations, dans les usines, à bord des
vaisseaux de l'Etat et dans les ménages prudents.

Les allumettes, depuis la guerre de 1870, ont
été frappées d'un impôt qui, d'après les clauses
du cahier des charges d'adjudication du Monopole,
représente, pour une consommation de quarante
milliards, quatre centimes par cent allumettes.
C'est 16 millions par an pour l'Etat!

Pour fabriquer quarante milliards d'allumettes en
bois, il faut employer 40,000 mètres cubes de
bois, ce qui représente, en volume, 400,000 hecto-
litres et, en longueur, un chapelet qui ferait 60
fois le tour de la terre ou plus de 6 fois le trajet
de la Terre à la Lune.

Illustration: The Story of the Match (d)

Doctors were not in agreement about the relative susceptibilities of men and women, the significance of the cough and general pulmonary congestion which affected all match workers, or the importance of preexisting dental caries. Dr. Alphonse Dupasquier found nothing more serious than chronic bronchitis in the Lyon factories in 1846 while Drs. Théophile Roussel and Sedillot in the same year found several Parisian workers who had contracted the disease.[8] Their only suggestions were preventive measures: cleanliness, ventilation, and the separation of workshops.

As a result of these reports, the Minister of Agriculture issued the first circular against this occupational hazard on October 15, 1852.[9] He required all the factory directors to provide physical separation between the workshops, ventilation, uninflammable building materials, and fire exits. However, there is no evidence of enforcement.

It was not consistent with the economic liberalism of the Second Empire to impose safety inspections on private indsutry or to hire professional inspectors. There was another obstacle to the enforcement of this decree: no one was sure of how many match manufactures there were or when they were in operation. Factories were usually small; they changed hands often. The advisory board of doctors on the *Conseil de Salubrité de la Seine* or Seine Health Council reported in 1858 that the number of match workers fluctuated "depending on the season, the state of commerce, the amount of exports."[10] There was a third reason why the Ministry's rules were not enforced: the casual attitude of owners and overseers even towards their own health. For example, many owners and overseers lived on the premises. In 1858 one such overseer and his wife were among the victims of necrosis although the wife had never worked in the factory.[11]

Despite the ineffectiveness of government supervision, there was one advance for public health and occupational health in 1856. A Swedish chemist named Lundstrom invented the safety match, which he made by putting potassium and sulphur on a match head and red phosphorous (also called amorphous phosphorous) on a separate striking surface. This match was safer because it was less likely to ignite by accidental friction, for example, when match boxes were jostled in the course of transportation. Thus, the risk of factory and warehouse fire was greatly reduced.[12]

In addition, red phosphorous did not provoke necrosis. The only workers who would be exposed to the poison would be those in chemical factories who processed white phosphorous into red. However, consumers did not appreciate these matches for two reasons: they were not accustomed to the inconvenience of using both hands to light a match and the red phosphorous was said to deteriorate with humidity. Hence, anyone who used matches in the fields, in forests, or near water would not buy them. For export to the tropics, they were especially undesireable.[13]

Nevertheless, five of the largest manufacturers petitioned the Minister of Agriculture to impose the exclusive use of red phosphorous. The Minister turned to the doctors on the *comité consultatif d'hygiène publique*, who issued a unanimous report in favor of a ban and substitution.[14] Indeed, by this time there had been a dozen necrosis cases in Lyon alone plus suicides and accidental poisonings by the ingestion of solutions made from match heads. The French holder of the patent on red phosphorous was willing to place it in the public domain. The sole French manufacturer of red phosphorous promised to cooperate with whatever decision the government arrived at. According to Dr. Ambroise Tardieu, the reporter for the *comité consultatif*, the cost of a kilo of red phosphorous matches might be ten centimes more than white phosphorous, but such a modest rise in price would not affect sales. Tardieu considered that France's export markets in South America and Australia were secure, seriously underestimating the competition that was to come from the German states, Sweden, and Japan. However, these suggestions fell on deaf ears. Neither the *comité consultatif d'hygiène publique* nor the Academy of Medicine, which also called for a ban in 1856, had any political power over occupational health.

As the industry expanded, an important factory opened in a northeast suburb of Paris: the Rimailho brothers' factory in Pantin. Although it was authorized by the Ministry of Agriculture, Commerce and Public Works in 1858, the factory was criticized by the doctors on the Seine Health Council for its deplorable lack of ventilation. The doctors proposed that its next authorization be refused.[15] Despite the Council's protest, the factory at Pantin received a renewed permit in 1868. If the government had used its authority to close that one factory, the entire history of the French struggle against white phosphorous necrosis might have been different because this plant, defective from the start, was to employ some of the most dedicated fighters against necrosis. Because of its location close to the seat of government and the major newspapers, this factory would be heard from in later years.

Unpleasant as it was to live near a match factory, it was worse to work there. Marx wrote to this effect in *Capital*:

> "the manufacture is on account of its unhealthiness and un-pleasantness in such bad odour that only the most miserable part of the labouring class, half-starved widows and so forth, deliver up their children to it,..."[16]

His impression of working conditions was confirmed by the French parliamentary inquiry of 1872 on the condition of the working class. In that report the workforce at the Aubervilliers match factory was described as eighty workers: half children, most illiterate.[17]

After the military defeat of 1870 and payment of reparations, the National Assembly taxed the match industry by the law of September 4, 1871. The results were disappointing: only 5,923,788 francs in 1872.[18] Two Marseille manufacturers suggested that the State declare a monopoly over the industry, similar to the one which had existed in the tobacco industry since 1730.[19] Competition from small manufacturers was especially fierce in the Rhône Valley and small manufacturers were more likely to evade taxes. The Marseille manufacturers hoped to be handsomely compensated for the buyout of their property. The government saw its interest in the consolidation of the industry in a few large factories.

Thus, for the next two decades the government tried the unhappy experiment of public ownership but private administration. At the request of the Ministry of Finances, Parliament established a monopoly by the law of August 2, 1872.[20] They also provided twenty million francs for the expropriation of all the existing factories. According to this law, the Ministry of Finances through the *Administration des Contributions Indirectes* would be the sole owner of all match factories with the exclusive right to produce, set standards and determine prices. However, rather than occupy its personnel with the day-to-day operations of a business, the Ministry of Finances drew up a twenty-year contract with a private company, the *Compagnie Générale des Allumettes de France*.[21] This contract could be terminated by either party after five, ten, or fifteen years. The Company agreed to make an annual payment of 16,030,000 francs. In addition, the company was to share with Finances any profits if the sales exceeded forty billion matches a year. In exchange for the right to that profit, the Company would produce only specified quantities and qualities of matches. An inspector from the Ministry of Finances was to sign receipts for all shipments which entered and left each factory. Authority over public health matters remained with the Ministry of Agriculture, Commerce, and Public Works.

It was not possible for the *Compagnie Générale* to provide management and keep records for hundreds of small workshops. The Ministry of Finances bought out all the match factories, continuing production in only eleven of the largest and terminating the jobs of as many as 20,000 workers.[22] (See Map I: Match Factories in 1875.) The years 1872-1875 were transitional because the owners of individual factories retained the right to produce and sell until they had received financial compensation.[23] Only by January 1, 1875 did the *Compagnie Générale* enjoy a *de jure* monopoly of production and sale.

The producers of contraband matches, of course, remained *de facto* competitors. An anti-contraband law was passed in 1875 but was revoked in 1876 due to popular pressure, especially in the Bouches-du-Rhône.[24] This area was never reconciled to the government monopoly. The Spanish border remained porous to smugglers and contraband continued to pass well into the 1890s.

However, the initial loss of as many as 20,000 jobs helped to change the legal French industry from one of part-time and seasonal workers in 1872 to one of full-time year-round workers by the 1880s. Still the *Compagnie Générale* was not willing to invest in improved buildings or medical attention for those workers. According to the Seine Public Health Council's report of 1872-1877, the Pantin factory was still not in compliance with ministerial regulations on health and safety.[25] Indeed, the *Compagnie Générale*'s representative to the *Conseil d'Etat* admitted the deficiency but argued that to do otherwise was impossible in that building:

"Application of the regulations would mean the reconstruction from top to bottom".[26]

The *Compagnie Générale* prevailed. The Pantin factory was neither rebuilt nor improved during the remainder of its lease. The first recorded worker protest on health and safety occurred at this factory in 1888 when the widow of a necrosis victim sued. Unfortunately, only *Le Temps* mentioned the lawsuit and did not indicate its outcome.[27]

From the first, the results of this collaboration were unsatisfactory to the Ministry of Finances. The *Compagnie Générale* reported that it had made no profits in its first five years of operation, 1872-1877, and had even had to use some of its capital to make its annual payments to

Finances. In the next five-year period, the *Compagnie Générale* paid dividends to its stockholders by cheating the public on the quality and quantity of matches.[28]

By 1883, the Ministry of Finances was sufficiently dissatisfied with the *Compagnie Générale* to consider the alternatives. The first one was to lease to another company. No other businessmen came forward. The second one was to make greater demands of the *Compagnie Générale*. This was done with modest results: by a new contract starting January 1, 1885, the amount paid to the State reached seventeen million francs for the first time.[29] However, the *Compagnie Générale* again paid dividends to its shareholders from 1885-1887 by withholding money from the Ministry of Finances.[30] The Ministry then considered its third alternative: to return the industry to the private sector. Small manufacturers who had not recovered after the buyout had continued to press for this. In Marseille, for example, they had convinced the City Council to vote resolutions to that effect in 1878 and in 1881.[31]

In April 1888, the Minister of Finances was Paul Peytral, who had been a radical deputy from Marseille since 1881. He was sympathetic to the desires of the ex-manufacturers in the Bouches-du-Rhône. Peytral notified the *Compagnie Générale* that on December 31, 1889, he intended to return match manufacturing to the private sector. However, on February 22, 1889, he was replaced as Minister of Finances by Maurice Rouvier. Rouvier was less impressed by the ex-match manufacturers than he was by the Director-General of the *Administration des Contributions Indirectes*. The Director-General advised against relinquishing control because it would cost ten million francs to buy out the stock of the *Compagnie Générale*. He assumed that the thirty-five million they had already spent on compensation to manufacturers was not recoverable.[32] Therefore, Rouvier proposed a fourth alternative: direct administration.

The question of public or private ownership was placed before the Chamber of Deputies in the last months of 1889. First, on November 19, 1889, Deputies Victor Leydet and Paul Peytral (Bouches-du-Rhône) introduced a bill to make match manufacturing free but subject to strict health standards.[33] To avoid the problem of tax collection, they proposed to go to the source by placing a heavy tax on white phosphorous itself. On November 21, 1889, Deputy Théophile David (Alpes-Maritimes), a doctor from a match-producing district, introduced an amendment to ban the use of white phosphorous. On

November 23, the bill was debated. Deputy Achille Werquan (Nord) added language on the safety of the workers. Deputy Jean-Baptiste Dumay (Seine), a socialist from a match-producing district, added an amendment against the employment of children under age sixteeen and another amendment in favor of the eight-hour day. The first passed; the second failed.[34] The Minister of Finances argued against the ban on the grounds that France would be unable to find an export market for any matches other than white phosphorous and that the French public would not buy them either. The bill failed.

Instead, the Ministry of Finances took over the direct administration by decree as of December 31, 1889. They closed all but six factories: Pantin and Aubervilliers (Seine), Bègles (Gironde), Trélazé (Maine-et-Loire), Saintines (Oise), and Marseille (Bouches-du-Rhône). (See Map II: Match Factories in 1890).

The Ministry expected to meet the criticism of all parties. They would clean up the industry to the satisfaction of the public and the workers. At the same time, they would raise the quality of matches and the quantity of revenue. After a brief closing of all the factories for inventory early in 1890, the Ministry of Finances rehired the 2000 workers and supervisory staff of the *Compagnie Générale*. Production was slower in the first year of direct administration than in the last year under the *Compagnie Générale*. In 1889 the *Compagnie Générale* had produced almost twenty-eight billion matches; in 1890 the *Direction Générale des Manufactures de l'Etat* produced less than fourteen billion matches.[35] However, by 1891 the French match industry had regained its previous production level. The transition was complete.

Map I: Match Factories in 1875

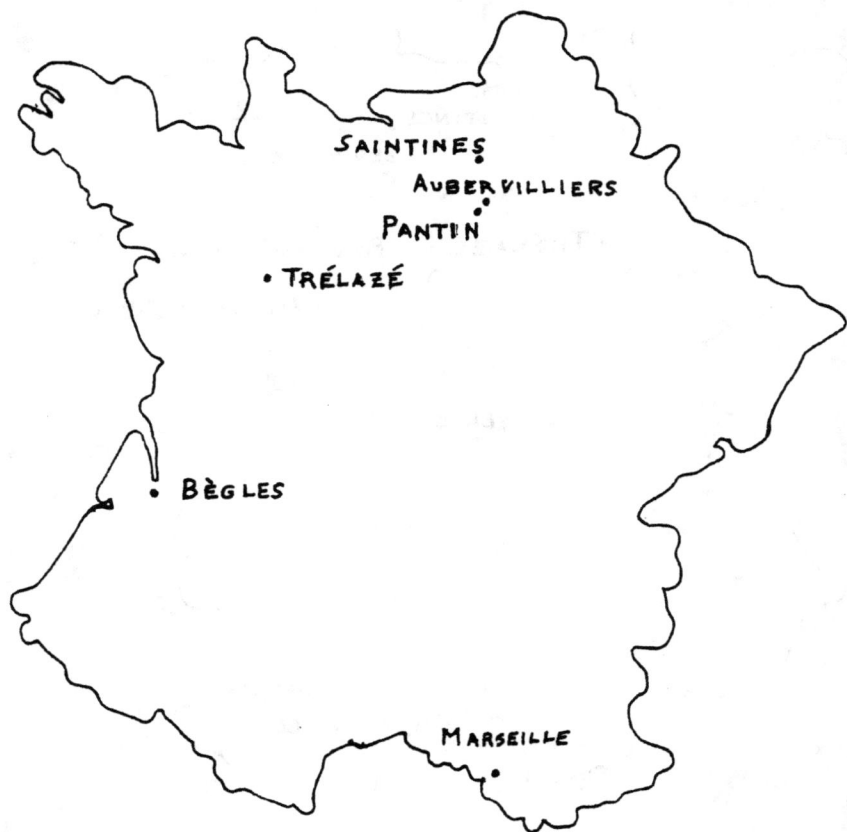

Map II: Match Factories in 1890

ENDNOTES CHAPTER II

1. Kammerer invented matches in Wurtemburg in 1832. Anne-Marie Clèdes, *Histoire de l'Allumette* , Paris: Service d'Exploitation de l'Industrie des Tabacs et des Allumettes, SEITA, n.d., 21.
2. idem, 24-5.
3. *Bottin. Annuaire-Almanach du Commerce et de l'Industrie*, Paris: Firmin-Didot, 1836, 4, 37.
4. ibid., 1865, 604.
5. M. Derme, *le Monopole des Allumettes*, Paris: Rousseau, 1911, 8.
6. Archives Nationales (AN) F22 528 Bureau de l'Association internationale pour la Protection Légale des Travailleurs, "Mémoire explicatif sur l'interdiction de l'emploi du phosphore blanc dans l'industrie des allumettes", 1904, 4-6.
7. Ambrose Tardieu, *Etude historique et médico-légale sur la fabrication et l'emploi des allumettes chimiques*, Paris: J.B. Baillière, 1856, 17. Phosphorism is not recognized as an occupational disease today. Necrosis is.
8. Dr.Alphonse Dupasquier, "Mémoire relatif aux effets des émanations phosphorées sur les ouvriers employés dans les fabriques de phosphore et les ateliers ou l'on prépare les allumettes chimiques", *Annales d'hygiène publique et de médecine légale*, Paris: J.-B. Baillière, October 1846, 342-356; Dr. Sedillot, "Observations de nécroses des os de la face et d'affections pulmonaires survenues à des ouvriers employés à la fabrication des allumettes chimiques", *Comptes rendus hebdomadaires des séances de l'Académie des sciences*, Paris: Bachelier, 1846, 437; Dr. Théophile Roussel, "Sur les maladies des ouvriers employés dans les fabriques d'allumettes chimiques, et, sur les mésures hygieniques et administratives nécessaires pour assainir cette industrie", ibid., 292-295.
9. *Rapport général sur les travaux du Conseil de Salubrité de la Seine, 1872-1877*, Paris: Chaix, 1877, 760.
10. ibid., 1858, 63.
11. idem
12. J.E. Lundstrom received a first class medal for his invention at the Paris World's Fair of 1856. Alphonse Chevallier, "Mémoire sur les allumettes chimiques", *Annales d'hygiène publique*, Volume XV, 1861, 330.
13. The French match industry relied heavily on exports in the 1850s and 1860s. The Roche factory in Marseille made two-thirds of its profits abroad. Derme, ibid., 8.

14. Dr. Ambroise Tardieu, *Etude*, 17. Physicians' organizations voted for a ban on white phosphorous seven times between 1856-1895 within the *conseil d'hygiène de la Seine*, the *comité consultatif d'hygiène publique*, and the Academy of Medicine. None of these bodies had political power. The first two were advisory bodies of doctors with an interest in public health. The *conseil d'hygiène* served at the request of the police. The *comité consultatif d'hygiène publique* depended on the Public Assistance Administration, which reported to the Ministry of Commerce. The Academy of Medicine represented the luminaries of the medical profession chosen by their peers. Their reports were widely read within the medical profession, but again, their decisions were not binding on any branch of the government.

15. *Rapport Général Conseil de Salubrité*, 1858, 63.

16. Karl Marx, *Capital*, Volume I, New York: International Publishers, 1974, 246.

17. Dr. Alphonse Dupasquier, "Mémoire," 354; Dr. Ambrose Tardieu, *Etude*, 14; Archives de la Préfecture de Police de la Seine (APP), *Enquéte parlementaire des Classes Ouvrières, 1872*, B A/400. The Aubervilliers factory was first authorized in 1867.

18. M. Derme, ibid., 8; Direction Générale des Manufactures d'Etat, *Compte en matières et en deniers de l'exploitation du monopole des allumettes chimiques*, Paris: Imprimerie nationale, 1898, 86-87.

19. The Marseille manufacturers were Roche and Caussemille. Derme ibid., 121.

20. Derme, idem.

21. The owners of the *Compagnie Générale* were Misters Pillet, Will, Vignal, and Archdeacon. They raised capital by selling 20,000 shares of stock at 500 francs each. Derme, ibid., 19.

22. In Marseille (Bouches-du-Rhône), they kept the factory of Roche and Caussemille and the factory of Four and Germain; in Nantes (Loire-Atlantique), Toyon-Delpit; in Trélazé (Maine-et-Loire), Lebatteux; in Blenod-les-ponts-à-Mousson (Meurthe-et-Moselle), Zeigler; in Orthez (Pyrénées-Atlantique), the Piétrat factory and the Seris factory; at Chalons-sur-Saóne (l'Yonne), the Crozet-David factory; at Pantin (Seine), the Rimailho factory; at Aubervilliers (Seine), the Delabarre factory; at Bellac (Haute-Vienne), the Sautrot factory. The Crépu-Pariset factory at Saintines (Seine-et-Oise) was closed by the Ministry of Finances in

1874 but reopened in 1876. Derme, ibid., 27; La Société nationale d'exploitation industrielle des tabacs et allumettes (SEITA), "Les allumettes", pamphlet 1982, *Le Sémaphore de Marseille*, May 18, 1889.

23. Some manufacturers took advantage by deliberately overproducing in order to prolong their time of sales or to demand a higher price at buyout. There were legal suits over the price of several factories. For example, the Ministry of Finances offered Roche and Caussemille 1,100,000 francs. Roche and Caussemille secured a court decision that the property was worth 5,900,000 francs. In total the Ministry of Finances spent not 20,000,000 francs, as originally allocated, but 34,862,463 francs on land, buildings, equipment, stock, and legal fees. Derme, ibid., 23. Roche and Caussemille then opened factories in Ghent, Belgium; Algiers and Bona, Algeria; Turin and Probesi, Italy. Great Britain, Parliamentary Papers, Thomas E. Thorpe, *Use of phosphorous in the manufacture of lucifer matches*, London: Eyre and Spottiswoode, 1899, 484.

24. The law of January 28, 1875 provided for fines of 300-1000 francs for the possession of contraband matches. A more controversial law was passed on July 28, 1875. It permitted the *Compagnie Générale* to hire anti-contraband agents and empowered them to make house searches. Furthermore, by the decree of August 10, 1875, illegal producers and vendors could be arrested. "In a few months, 20,000 house searches were made. Entire communes were searched house by house. Even the most respectable people did not escape the investigations." Five thousand violators were discovered. There were protests in the press. The law was revoked in July, 1876. Léon Say, a staunch foe of government interference in business, was the Minister of Finances for most of this time. Derme, ibid., 40.

25. *Rapport général Conseil de salubrité 1872-1877*, 760.

26. *ibid., 1878-1880*, 820.

27. *Le Temps*, November 5, 1888.

28. According to Deputy Victor Leydet (Bouches-du-Rhone), he had found only 90 matches in a box which was labelled 150 matches. Of these 90, fifty did not light. Derme, ibid., 45.

29. New rules which were issued on July 7, 1884, specified that the Ministry of Finances would impose its accounting methods and verify all accounts. In addition, to prevent the smuggling of

white phosphorous from the factories, the *Compagnie Générale* was to provide an apartment on site to State inspectors. The apartment was to have a view of the only entrance and exit. All other openings were to be covered with bars. If possible, the entire factory was to be surrounded by a circular walk. France, Ministère des Finances, *Cahier des Charges pour l'exploitation du monopole des allumettes chimiques (le 7 juillet 1884)*, article 7, Paris: Chaix, 1884. The new contract of January 1, 1885 was to encourage productivity by giving the *Compagnie Générale* one hundred per cent of its profit from exports instead of ninety percent. In addition, the *Compagnie Générale* could keep all its profits from the domestic sales after the first thirty-five billion matches instead of the first forty billion. The annual payment to the Ministry of Finances was raised to 17,010,000 francs. Although this was a 20-year contract, it could be annulled by either party on one year's notice. Great Britain, Parliamentary Papers, *Use of Phosphorous*, 680.

30. Derme, ibid., 46.
31. Marseille, Conseil municipal, *Déliberations, 1878*, Marseille: Moullet, 1879, 443 and *Déliberations, 1881*, 251.
32. Derme, ibid., 47.
33. Co-sponsors of the bill of November 19, 1889: Deputies Victor Leydet, Paul Peytral, Camille Pelletan, Félix Granet, and Antide Boyer from the Bouches-du-Rhône; Alexandre Millerand and Tony Révillon from the Seine; Jean Lachize from the Rhône. Journal Officiel, Chambre, *Débats*, 88.
34. On the eight-hour day, the vote was 192 in favor and 263 opposed. That was a considerable minority in favor of such an advanced demand in 1889. Journal Officiel, Chambre, *Débats*, November 23, 1889, 132.
35. The private company produced twenty-two to twenty-nine billion matches almost every year between 1873 and 1889. The profit was sixteen to seventeen million francs each year. The Ministry of Finances produced the same number of matches from 1891-1894 but the profit was nineteen to twenty million francs. Direction Générale des Manufactures de l'Etat, *Compte en matières et en deniers*, 86-89.

CHAPTER III: DEMOGRAPHY:
A PROFILE OF THE MATCH WORKERS

In this chapter, the censuses of 1891 and 1896, personnel dossiers, and information from the statistical annuals published by the Ministry of Commerce are used to test several theories about the effects of the match workers' material conditions on their behavior as organized workers. Whenever possible, they are compared to their neighbors in the commune of Aubervilliers or to the French population as a whole.

I have used information from the handwritten censuses of 1891 and 1896 on 144 male match workers and 258 female match workers in the communes of Pantin and Aubervilliers. The censuses provided the address, names of all the individuals who lived at that address, their ages, their occupations, and their relationships to one another. This data was supplemented with the personnel dossiers of 100 of these workers. For each one, some or all of the following information was found: name(s), year of birth, date(s) of entry into the factory, date at which the worker received a permanent job (*titulaire*), the dates and occasionally the reasons for absences, the number of days worked from 1892-1904, the date of receipt of pension, the reason for retirement, the date of death, and the names and birthdates of pension recipients.[1]

Let us begin by asking what sort of people lived in the communes of Pantin and Aubervilliers. These were working class suburbs northeast of the city limits. Small manufactures grew there or relocated there from the center of Paris after 1860. By 1886 the neighborhood was five-sixths family units and one-sixth individuals who lived alone, usually young single people. By 1891 it was six-sevenths family units. The families were stable in size from the 1886 census to the 1896 census. Half of the married couples had one or two children. It was a neighborhood of young families in formation.

The match workers were similar to their neighbors in every way but one. Because the Ministry of Finances named 700 of them as permanent workers at the Pantin-Aubervilliers plants in the early months of 1890, most of the match workers who were found in the 1891 census were still match workers at the time of the 1896 census. Of course they were five years older, so the match worker population aged slightly in comparison with the neighborhood as a whole between 1891 and 1896. That is, the percentage of 16- to 21-year old match workers was much lower than the percentage of 16- to 21-year olds in the neighborhood. As might also be expected, the percentage of married match workers and parent match workers increased through the 1890s. In contrast with their neighbors, match workers' families grew.

When we examine the match workers, we are especially interested in four questions about how their personal circumstances may have affected their militance on the job. First, Karl Marx's description of the match workers of the 1860s will be tested against the information from the 1890s to learn whether the match workers were particularly militant because they were indeed desperate widows and orphans. Second, we compare their household sizes and family structures to their neighbors' in an effort to understand whether anything in the match workers' home lives made them different from their less militant neighbors. Third, the census and personnel dossier information shed light on the health of the Pantin-Aubervilliers match workers. Given the weakness allegedly provoked by necrosis, it is worthwhile to look at the job stability of the match workers. Next, their child-bearing patterns are examined for evidence of any inability to produce healthy offspring as alleged by the Federation and individual match workers.

Were the match workers who began to work under the private company between 1872-1889 children from the most miserable ranks of society? Although the personnel dossiers do not indicate the occupations of the match workers' parents, information was available on the ages at which 66 of the workers entered the Pantin-Aubervilliers match factories. Indeed, some (11%) began to work before their sixteenth birthday. However, most of them (50%) began to work between the ages of 16 and 20. The average starting age was almost 22. This indicates that the match industry was not or was no longer a refuge for child labor in the 1870s and 1880s. Although the match factories hired many young people, as might be expected in any job which required only speed and manual dexterity, they did not hire many children. (See Chart I: Age at Which Workers Entered the Match Factories 1872-1889).

By 1891 the age distribution showed very few workers under the age of 16: no males and only six females (3% of the total). The majority of the workers were in their prime working years: aged 16 to 30. (See Chart II: Match Workers by Age in 1891).

CHART I: AGE AT WHICH WORKERS ENTERED
THE MATCH FACTORIES 1872-1889

Age	Number of individuals	%
11-15	7	11%
16-18	21	32%
19-20	12	18%
21-25	12	18.5%
26-30	8	12.5%
31-35	2	3%
36-40	2	3%
41-45	0	0
46-50	0	0
51-55	2	2%

n = 66

Average age 21.7

CHART II: MATCH WORKERS BY AGE IN 1891

Age	Number of individuals	% of individuals
0-16	6	3%
17-19	27	15%
20-24	50	27%
25-29	37	20%
30-34	21	11%
35-39	13	7%
40-44	12	6%
45-49	11	6%
50-54	3	2%
55-59	4	2%
60 +	3	2%
Totals	187	101%

Did the composition of their households suggest the financial desperation alluded to by Marx? To answer this question, we must keep in mind that the average wage in the *département* of the Seine in 1891 was 6 francs 20 centimes per day for men and 3 francs 15 centimes for women. Assuming at least a ten-hour day, the average male manual laborer earned from five francs a day, if he was an unskilled general laborer *(homme de peine)* to eight francs a day if he was a roofer or a carpenter. In comparison, the match workers earned moderate wages: an average of 4 francs 70 for the men and 3 francs 25 for the women in 1893. However, their job was especially desirable. For unskilled women in particular, it was an excellent job. First of all, it was thought to be secure. The employer would not go out of business; a permanent employee was not fired for anything less than theft or insubordination; it was not seasonal. In addition, there were bonuses for childbirth or military service and a pension after thirty years of work. Unlike their neighbors, match workers could plan their futures.[2]

Indeed, the compositions of the match workers' households indicate some careful decision-making to avoid poverty and even to establish lives of modest comfort. (See Chart III: Working Members and Dependents in Match Workers' Households in 1891). Eleven to fourteen percent of the Pantin-Aubervilliers match workers lived alone and supported only themselves; the women in these situations probably suffered financial hardship. Approximately four of every ten match workers lived with a working spouse or boyfriend/girlfriend but no dependents. Typical of these were Anaise Andrieux, a 21-year old match worker and her 21-year old boyfriend, a butcher's assistant. These people were relatively comfortable. However, in more than a quarter of the match workers' households there were three or more wage earners supporting children and the occasional elderly parent. These households may have been needy from time to time depending on the number of non-working members, the stability of the working members' jobs, and the amount of their earnings. For example, the Leprosnier family was fairly secure with a 47-year old husband who worked as a fitter, a 47-year old wife who worked at one of the match factories, a 19-year old stepdaughter who was a daily operative at a toy factory, and a 6-year old son. Even if the 19-year old woman left to form her own household, this family would manage unless the man of the house fell ill or died.

In contrast, there were several households in which there was no adult male wage earner with a good job but rather a mother and grown children. In these cases, we see that indeed the match factories still lived up to their reputation as the refuge of widows and orphans. The sons and daughters in these families had to start work young at unskilled jobs. They could not establish their own households without plunging the mother into poverty or making sure that another sibling would take up the slack. Let us take the case of Marie and Marguerite

CHART III: WORKING MEMBERS AND DEPENDENTS IN MATCH WORKERS' HOUSEHOLDS IN 1891

Number of households	Number of dependents	Percentage households	Working members
Male matchworkers			
7	-	11%	One-person household
16	15	24%	Husband and wife
9	3	14%	Unmarried couple
12	30	18%	Husband or wife only
1	1	2%	Single man or woman only
8	25	12%	Father and child(ren)
4	2	6%	Mother and child(ren)
5	4	8%	Husband, wife, and child(ren)
4	6	6%	Other
66	6	101%	Totals
Female matchworkers			
21	-	15%	One-person household
34	29	25%	Husband and wife
24	11	17%	Unmarried couple
-	-	-	Husband or wife only
-	-	-	1 friend only
15	44	11%	Father and child(ren)
13	10	9%	Mother and child(ren)
16	25	12%	Husband, wife, and child(ren)
15	19	11%	Other
146	148	100%	Totals

Average number of dependents in households of male match workers 1.3
Average number of dependents in households of female match workers 1.0

Landour. Their 49-year old mother worked as a cook, a poorly paid job. Twenty-three year old Marie and 19-year old Marguerite had been match workers since the age of 15. One of them had a year-old son. Personnel dossiers reveal that they both married eventually, although the dates of their marriages are not indicated. At the time of the 1891 census, the fifth member of their household, their 21-year old brother was working as a packer (*emballeur*) which was an unskilled, poorly paid job. For Marie and Marguerite to marry, which would presumably lead to new dependents, probably required either a heavier burden on the son, the death of the mother, or the acceptance of the mother's solitary poverty.[3]

In the remaining quarter of the match workers' homes there was an important difference between male and female match workers. Two of every ten male match workers were able to support a non-working woman with or without children. These one-breadwinner households had to live modestly but were not miserable. In fact, it seems that as male workers gained seniority, they became increasingly able to support a dependent woman. The average childless husband with a working wife was 28 years old; the average husband/father with a working wife was 30 1/2 years old. The average childless husband whose wife did not work was 31; his counterpart with children was almost 34.[4]

However, only 6% of the women were able to be the sole support of their dependents. Typical of them was the household of Marie Jouhaux, who, at the age of 24, supported four daughters aged seven, six, four, and two.[5] Some of the women whom we might expect to find living with their children but no other adults were found instead among the 10% who were classified as "other": living with a woman friend, a cousin, or an in-law. In addition to the lower wages of the women, these households reflected the social constraints on women to find a male protector or at least another adult of either sex to help with the children.

When we discuss the other employed members of the households, we should realize that the most frequent place of employment for any member, whether male or female, single or married, was the match factories (33%). Forty-five percent of the employed household members of male match workers and 25% of the members of female match workers' households were also employed there. The other types of employment were a mixture of manufacturing (38%), the building trades (11%), service (10%), and sales (8%). These job categories confirm that the match workers of the 1890s issued from families of unskilled workers.

In size, the match workers' households looked much like those of their non-match worker neighbors. (See Chart IV: Comparison of Number of Household Members in the Commune of Aubervilliers and in Match Workers' Households in 1891). In the entire commune of Aubervilliers the average number of people per household was 3.2; among the match workers the average was the same. The percentage of individuals who lived alone, who lived with one other, two others, etc. was similar in both cases. We conclude, therefore, that by 1891 the match workers were typical Pantin-Aubervilliers residents rather than the youngest or the neediest of workers.

CHART IV: COMPARISON OF THE NUMBERS OF HOUSEHOLD MEMBERS IN THE COMMUNE OF AUBERVILLIERS AND IN THE HOUSEHOLDS OF THE MATCH WORKERS IN 1891

	Aubervilliers		Match Workers	
Households of	Number	Percentage	Number	Percentage
one	1421	19%	28	14%
two	1854	24%	55	28%
three	1463	19%	46	23%
four	1204	16%	24	12%
five	801	10%	22	11%
six	513	7%	9	5%
seven or more	413	5%	14	7%
Totals	7669	100%	198	100%
Averages	3.2 per household		3.2 per household	

Data from AD Paris, *Recensement 1891, Département de la Seine, Arrondissement de St. Denis, Canton de St. Denis, Commune d'Aubervilliers*

Yet they claimed that phosphorism was a general debilitating condition which they all endured. "Some of us have it in the brain, others in the belly."[6] If this was so, we might expect to find irregular work patterns with frequent interruptions. That is, many might have left the match factories when they became too ill, too weak, or too frightened to stay but before they were officially classified as necrosis victims. If we look at those match workers of 1891 whose place of employment in 1896 is also known, we find that very few moved into other jobs (9%) or became fulltime homemakers (5%).[7]

On the other hand, if we look at those match workers of 1896 whose place of employment in 1891 is also known, we see that few moved in from other jobs (19%).[8] This lack of mobility after 1891 suggests that the factories had hired enough permanent workers in the first months after the government takeover to have their core staff. Those who were hired afterwards were primarily young people who already had a relative at the factories to help place them.

It seems then that a job at the match factories was a desirable one: once a man or a woman had a permanent place, he or she would stay despite the real or imaginary threat to health. Indeed even reporters who were sympathetic to the match workers' health wrote that some of the workers did not take necrosis as a serious threat. Rather, once they had evidence that they were afflicted, they were likely to continue to work in the hopes of an early pension.[9]

However, the Federation claimed throughout the 1890s that their members suffered from one more health hazard: reproductive difficulties. In her 1897 series on working women, Aline Valette wrote in the feminist newspaper, *La Fronde*, that necrosis caused not only the rotting of teeth and bones but also miscarriages. She accused the Ministry of Finances of killing pregnant women along with their unborn children.[10] Although current medical manuals do not list reproductive hazards among the effects of white phosphorous, there has been no incentive to do systematic research on the question since 1906 when most match manufacturing countries banned the poison.[11] At that time the French, ill-informed as they were, were the most knowledgeable. If the match workers were correct in their accusations, the evidence might not be apparent from their decisions to retire or to take other work; they may have judged that the wages and job security at the match factories were more important than childbearing. However, reproductive difficulties could be detected from the census in an abnormally high proportion of childless couples. Of coursea we must take into account that the match workers and their household members were young in comparison with the French population as a whole (See Chart V: French Population and Match Workers' Households by Age and Sex in 1891). They clustered in the ten-year bracket from age 20-29: the childbearing years. There were very few match workers or household members over the age of 50, which makes impossible a comparison of their birthrate to the national birthrate per 100 individuals in the population. It is surprising, therefore, that they had the same proportion of household members under the age of 15; we would expect more children in the households of these people in their prime reproductive years. Did this discrepancy reflect reproductive difficulties or other social factors?

CHART V: FRENCH POPULATION AND MATCH WORKERS' HOUSEHOLDS BY AGE AND SEX IN 1891

Age	French Males	French Females	Households of Male Match Workers	Households of Female Match Workers
0-4	9%	8.5%	10%	10%
5-9	9%	9%	10%	10%
10-14	9%	8.5%	6%	10%
15-19	9%	8.5%	10%	12%
20-24	8%	9%	14%	16%
25-29	8%	7.5%	19%	11%
30-34	7%	7%	8%	5.5%
35-39	7%	6.5%	6%	7%
40-49	12%	12%	11%	14.5%
50-59	10%	10%	3.5%	4%
60-69	7%	7.5%	2%	1%
70-79	4%	4%	1%	-
80-89	-	1%	-	-

62 households of male match workers = 149 individuals
152 households of female match workers = 495 individuals
Source: Archives départementales de la Seine, D2M8 Dénombrement 1891, Article 3 Aubervilliers, Article 25 Orly-Pantin

To answer this question we must evaluate the meaning of marriage to working class people in these years. Although there was not much social stigma on free unions and the influence of the Church was weak, most working class couples formalized their relationships eventually. The census data suggest that the decision to have or legitimize a child had much to do with the decision to marry. First of all, women match workers married later than the average Frenchwoman. (See Chart VI: Frenchwomen and Women Match Workers by Age and Marital Status in 1891 and 1896). Perhaps the fact that they could earn their own keep prevented them from making hasty marriages just to provide for themselves or to remove a burden from their parents. Second, those who lived with a boyfriend were on the average younger than those who were married. (See Chart VII: Ages of Married Match Workers and Unmarried Cohabiting Couples in 1891). The average age of boyfriends was 27.9 while the average age of husbands was 32.5; the average age of girlfriends was 25.9 while the average wife was 32.8 years old. Most of the match workers who had children with their live-in girlfriend or boyfriend at the time of the 1891 census had married by the time of the next census.[12] It seems that living together was not a symptom of the disaggregation of the working class family under the impact of industry; nor was it the refuge of the poor woman who could not support herself on her own wages. Rather, it was a testing period from the ages of 19 to 29 after which the two people usually married.

Most of the girlfriend-mothers gave birth between the ages of 19 and 24 while most of the married mothers gave birth between 25 and 34. This suggests the possibility that some of the girlfriends became pregnant either by accident or in an effort to make the child's father marry them while more of the married women became pregnant deliberately. The age discrepancy may also reflect the fact that most of the match workers started their families in their late twenties. Therefore, perhaps these figures indicate that when a girlfriend in her late twenties became pregnant, her boyfriend was likely to marry her by the time the censustaker next came around, and none of the information sources available to this researcher would indicate that the child had been born out of wedlock.

CHART VI: FRENCHWOMEN AND WOMEN MATCH WORKERS BY AGE AND MARITAL STATUS IN 1891 AND 1896

Age	Single	Married	Divorced/Widowed	
15-19	95%	5%	-	All Frenchwomen 1891
	96%	4%	-	Match workers 1891
	84%	16%	-	Match Workers 1896
20-24	62%	37%	1%	All Frenchwomen 1891
	85%	10%	4%	Match Workers 1891
	71%	27%	2%	Match Workers 1896
25-29	32%	65%	3%	All Frenchwomen 1891
	43%	57%	-	Match Workers 1891
	56%	43%	-	Match Workers 1896
30-34	21%	73%	5%	All Frenchwomen 1891
	36%	43%	21%	Match Workers 1891
	15%	69%	15%	Match Workers 1896

1891 109 matchworkers
1896 113 matchworkers

CHART VII: AGES OF MARRIED MATCH WORKERS AND UNMARRIED COHABITING COUPLES IN 1891

Age	Husbands	Single men	Wives	Single women
Under 20	-	-	2.7%	19.2%
20-24	3.7%	30%	13.5%	46.2%
25-29	33.3%	60%	32.4%	11.5%
30-34	37%	-	16.2%	3.9%
35-39	11.1%	-	13.5%	3.9%
40-44	3.7%	-	10.8%	7.7%
45-49	7.4%	-	2.7%	7.7%
50-54	3.7%	10%	2.7%	-
55-59	-	-	2.7%	-
60-65	-	-	2.7%	-
Average age	32.5	27.9	32.8	25.9
Number of individuals	27	10	37	26

Next, we must look at the match workers with and without children. The ages of parents versus non-parents further confirms the hypothesis that child-bearing for male and female match workers was more related to nearing the age of 30 than to any other variable. (See Chart VIII: Ages of Match Worker Parents and Non-Parents by Age in 1891). The typical childless boyfriend or husband was 30.7 years old. His male counterpart with one or more children was only one year older (31.5). Among the women, the age gap was slightly wider. The average childless girlfriend or wife was 26.8 years old. The average mother was 32.9. This six-year gap indicates that many 19- to 29-year old female match workers who lived with a man did not give birth despite years of sexual intimacy. Was their childless state a choice or the result of white phosphorous intoxication?

CHART VIII: AGES OF MATCH WORKER PARENTS AND NON-PARENTS AMONG COHABITING COUPLES IN 1891

Age	Childless		Fathers	
	Husbands	Single men	Husbands	Single men
Under 20	-	-	-	-
20-24	-	2	1	1
25-29	4	4	5	2
30-34	2	-	8	-
35-39	1	-	1	-
40-44	-	-	1	-
45-49	-	-	2	-
50	1	1	-	-
Average	32.3	28.9	32.5	25.7
		30.7		31.5

Age	Childless		Mothers	
	Wives	Single women	Wives	Single women
Under 20	-	4	1	1
20-24	5	5	1	5
25-29	5	1	6	2
30-34	4	-	2	1
35-39	1	1	4	-
40-44	1	1	3	1
45-49	-	-	1	2
50	-	-	2	-
Average	29.1	23.7	35.4	28.8
		26.8		32.9

The evidence suggests a choice., (See Chart IX Part I: Married Match Workers and Cohabiting Couples Aged 19-29 by Number of Children in 1891). Only one of every three households of match worker boyfriends and girlfriends aged 19 to 29 had a child. Among married match workers of the same ages, more than half of the households had a child; one quarter of these households of young married match workers had two children. If white phosphorous had affected the women's reproductive abilities but not the men's, there would have been a discrepancy between the households of female and male match workers. Let us reexamine the households of only those whose sexual partners did not work at the match factories. (See Chart IX Part II). Indeed, there is a discrepancy but it is not between the sexes. Only one of five of the unmarried male match workers aged 19-29 with a live-in girlfriend was a father. Three of four of the married male match workers of the same ages had a child. Almost half of the unmarried female match workers with live-in boyfriends had a child while almost a third of the married female match workers had two. The fact that women match workers who had not formalized their relationships by marriage had a lower birthrate than their sisters (in several cases, the women under consideration were literally sisters) and that the boyfriends had even fewer children points to a choice rather than a disability, a refusal to reproduce rather than industrial poisoning.

CHART IX PART I: MARRIED MATCH WORKERS AND COHABITING COUPLES AGED 19-29 BY NUMBER OF CHILDREN IN 1891

Number of children	0	1	2	
Husbands	40%	50%	10%	n=10
Boyfriends	82%	18%	-	n=11
Wives	47%	26%	26%	n=11
Girlfriends	61%	39%	-	n=18
All Married Match Workers	45%	34%	21%	n=29
All Unmarried Cohabiting Match Workers	69%	31%	-	n=29

CHART IX PART II: MARRIED MATCH WORKERS AND COHABITING COUPLES WHOSE PARTNER DID NOT WORK IN MATCH FACTORIES BY NUMBER OF CHILDREN IN 1891

Number of children	0	1	2	
Husbands	25%	75%	-	n=4
Boyfriends	80%	20%	-	n=5
Wives	55%	18%	27%	n=11
Girlfriends	58%	42%	-	n=12
All Married Match Workers	47%	33%	20%	n=15
All Unmarried Cohabiting Match Workers	65%	35%	-	n=17

Next we notice that among match workers over the age of 30, none of the boyfriends was a father and only four of the girlfriends were mothers. The information on all four of the women makes them exceptional in one way or another.[13] This confirms the hypothesis that after the age of 30 those match workers who wanted children deliberately set about reproducing and that one step in that process was to marry. Indeed, some of the married mothers and fathers over the age of 35 had toddlers.[14] It seems that the marriage contract for the match workers was an announcement of their intention to reproduce as if, in reality, they did not fear the effects of white phosphorous.

How then do we explain that 35% of the married match workers had no children in 1891 (33% in 1896) while only 24% of married Parisians and 17% of residents of France in general were childless? And how do we explain that only five to six percent of the match workers' households had four or more children compared to ten percent in the Seine and 19% nationwide? (See Chart X: Households of Married Match Workers, Married Parisians, and Married Residents of France by Numbers of Children in 1891 and 1896). It may have been the relative youth of the match workers because most of them were under the age of 30 in 1891 and 1896. It may also have and Married Rbeen that their marriages were recent. After all, nationally, among people who had been married for five years or less, 32% had no children, 38% had one, 19% had two.[15] The sample of match workers who had been unmarried at the 1891 census but were married by the time of the 1896 census is too small to permit conclusions but the incidences of childless couples, parents of one, and parents of two is roughly comparable.[16]

Could it have been, then, that they had a high rate of miscarriages, stillbirths and infant mortality? Unfortunately, the handwritten census records did not indicate pregnancies, miscarriages and stillbirths. Although the annual statistical tables of births and deaths include stillbirths, these are not broken down by commune or by occupation. Therefore, no conclusion can be drawn. About infant mortality, however, it can be said that six of the seven match workers' children newborn to 36 months in 1891 were found in the 1896 census.[17] This is especially remarkable because 1892 and 1893 were epidemic years of cholerine, typhoid fever, and pneumonia. Tentatively, these survivors indicate that the match workers' children were not unduly subject to infant mortality.

CHART X: HOUSEHOLDS OF MARRIED MATCH WORKERS, MARRIED PARISIANS, AND MARRIED RESIDENTS OF FRANCE BY NUMBERS OF CHILDREN IN 1891 AND 1896

Number of children	France 1891	Seine 1891	Match Workers* 1891	Match Workers* 1896
Unknown	2%	9%	-	-
0	17%	24%	35%	33%
1	25%	27%	32%	28%
2	22%	20%	21%	24%
3	15%	10.5%	6%	10%
4	9%	5%	6%	4%
5	5%	2.5%	-	1%
6	3%	1%	-	-
7 or more	2%	1%	-	-
Average number of children	2.1	1.5	1.2	1.3

*The figures for the match workers do not include widows, widowers, divorcés.

What is more revealing about the well-being of their infants is that the number of newborns grew slowly but steadily every year from 1885 to 1896 with no fluctuations for strike years or epidemic years. In the midst of their press statements that their babies were dying, the match workers acted like vigorous forward-looking couples who waited until they had secure jobs, then deliberately started their families. (See Chart XI: Annual Number of Births to Match Workers 1885-1896 and Graph I: Births to Match Workers 1885-1896). The upward movement of the number of births does not jump after 1890 when the sanitary conditions of the factories were allegedly improved by the Ministry of Finances, the match workers' ability to purchase more and better food increased after the State raised their wages, or their hopes were raised once the State became their direct boss and the Federation became their intermediary. On the contrary, the increase in births was small but regular from 1885-1890, when their jobs were insecure and there was even discussion of returning the factories to the private sector. The increase reflects the fact that the match workers were young, healthy couples who were founding their families. The Ministry of Finances simply accentuated the tendency

by naming the workers of 1890 permanent so that there were fewer
new places for young single workers and strong incentives to stay.

CHART XI: ANNUAL NUMBER OF BIRTHS TO
MATCH WORKERS 1885-1896

Year	Number of births to matchworkers in 1891 census	Number of births to matchworkers in 1896 census
1885	5	1
1886	10	7
1887	10	7
1888	10	5
1889	9	8
1890	9	13
1891	11	18
1892	-	11
1893	-	15
1894	-	13
1895	-	21
1896	-	19
total matchworkers	208	236

It is possible to see a choice in the fact that so many of the match
workers were childless or considered their families complete after the
third child if we examine the ages at which match workers became
parents. (See Chart XII: Births by Ages of Fathers and Chart XIII:
Births by Ages of Mothers). Married male match workers fathered
more of their children than did the average Frenchman between the
ages of 20 and 29. After that age, their proportion of children dropped
off markedly. Among the women, the fertility pattern was almost
identical to the national pattern until the age of 30, at which point it
too dropped off markedly. This suggests that female and male match
workers were just as fertile as any other people living in France. We
can assume that the relative youth of the match worker fathers is proof
that when match workers wanted to have a child, white phosphorous
was no obstacle. Because the match workers had fewer worries about
money and job security than did many Frenchmen and women, those
who wanted to have children set about it a few years earlier than was
common in France. In fact, the age at which the match workers of
1891 had their last child further reinforces the hypothesis that male

match workers chose to have their children while they were young. (See Chart XIV: Match Worker Parents Aged 35-50 by Age at Last Birth in 1891 and 1896). The trend became more pronounced by 1896. To make a rough estimate of parents' ages at the birth of their last child, let us look only at those parents who were 35 years old or older in 1891 or 1896. Because in 1891 male match workers had only 29% of their children after the age of 35 and female match workers had eight percent of theirs, it can be assumed that the youngest child of many of these match workers at the 1891 census never had a younger sibling. In 1891 the average match worker father over the age of 35 had indeed just completed his family (at age 34.6). In 1896 the trend was unchanged; the average age at last birth was 33.9. For women, the tendency between 1891 and 1896 was to extend their childbearing years, which again belies any negative effects of white phosphorous. The average mother over the age of 35 in 1891 had a 3-year old as her youngest child; the average age at last birth was 31.6. By 1896 the average age at last birth was up to 36.9. This was because, in the heat of the anti-necrosis campaign, enough women match workers in their early forties had one more child to raise the average.

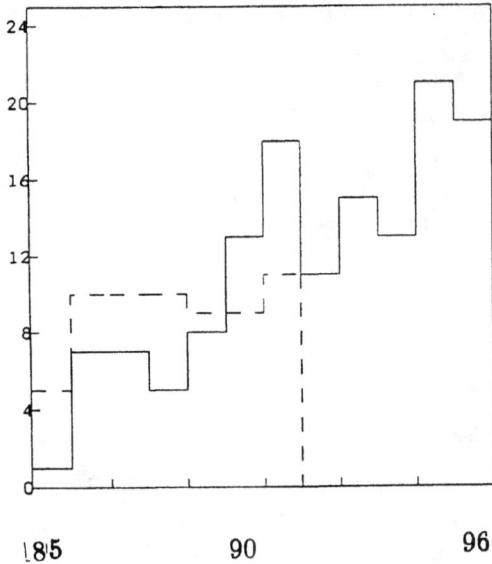

GRAPH I: Births to Match Workers 1885 - 1896
dashed 1891 census   solid 1896 census

CHART XII: BIRTHS ACCORDING TO AGES OF FATHERS: COMPARISON OF ALL MARRIED FRENCHMEN, MARRIED MATCH WORKERS 1886-1891, AND MARRIED MATCH WORKERS 1892-1896

Age of father	All married French	1886-1891 Married Match Workers	1892-1896 Married Match Workers
Under 20	0.4	-	-
20-24	5.7	17.0	10.0
25-29	27.0	40.0	51.0
30-34	27.7	11.4	24.0
35-39	19.3	8.6	10.0
40-44	11.7	14.3	2.5
45-49	5.4	5.7	2.5
50 or older	2.6	-	-
unknown	0.2	2.9	-

CHART XIII: BIRTHS ACCORDING TO AGES OF MOTHERS: COMPARISON OF ALL FRENCHWOMEN, MATCH WORKERS 1886-1891, MATCH WORKERS 1892-1896

Age	All French Women	Match Workers 1886-1891	Match Workers 1892-1896
Under 20	7.6	17.5	8.0
20-24	24.1	27.0	27.0
25-29	29.4	30.2	37.0
30-34	21.5	14.3	18.0
35-39	13.0	6.3	6.0
40-44	5.4	1.6	2.0
45-49	0.8	-	-
50 or older	0.1	-	-
unknown	0.3	3.2	2.0

We see that the material conditions of the match workers of the 1890s were far from those of the 1860s, although the Federation leadership knew how to play on the public's collective consciousness of this unskilled majority female trade as the refuge of widows and orphans. We have seen that indeed, in their family relationships, they lived like any of their working class neighbors. In general job stability, there is no evidence that phosphorism caused frequent absences from work or changes of jobs. In particular, for their rallying cry of reproductive hazards, there is no evidence.

What has emerged instead from this statistical analysis is a picture of the match workers as careful planners of their lives, working class people who intended to improve their own and their children's welfare. They usually waited until their mid-twenties to marry and their late twenties to have children. They stopped adding to their families when they had the desired number of children. They sometimes arranged their work choices so that the woman of the house could become a fulltime homemaker. They tried to place their children at the matchworks. All of this has implications for their trade union militancy. Perhaps they burst into union activity after 1890 not only because the State takeover made it possible but because they were families fighting for the welfare of their own parents, spouses, lovers, and children.

CHART XIV: MATCH WORKER PARENTS AGED 35-50
BY AGE AT LAST BIRTH 1891 AND 1896

	1891	1896
Average age at last birth		
Mothers	31.6	36.9
Fathers	34.6	33.9
Average age at time of census		
Mothers	40.4	40.6
Fathers	41.6	37.8
Average number of years since last birth		
Mothers	8.9	3.7
Fathers	7.0	5.0

10 mothers in 1891
9 mothers in 1896
5 fathers in 1891
9 fathers in 1896

ENDNOTES TO CHAPTER III

1. Archives Départementales de la Seine (AD Seine), D2M8 *Dénombrement 1891*, Article 3 Aubervilliers, Article 25 Orly-Pantin; *Dénombrement 1896*, Article 39 Aubervilliers; Archives at the Saintines (Oise) match factory, *Registre matricule pour l'inscription des ouvriers et ouvrières temporaires à dater du premier janvier 1909*, number 48, Manufacture de Pantin-Aubervilliers; Ministre du Commerce, *Annuaire Statistique de la France* (hereafter called *Annuaire Statistique*), Paris: Imprimerie Nationale, 1893-1934. Although most of the data is based only on Pantin-Aubervilliers workers, contemporary observors described them as respresentative of the French match workers as a whole in sex and age distribution. We know that there was a pattern of worker emigration to the twentiethth *arrondissement* of Paris for jobs at the match factories by provincial match workers. For these reasons we are assuming that the Pantin-Aubervilliers workers were also representative in family structure, child-bearing attitudes and general health.

2. *Annuaire Statistique*, ibid. 1893, Volume 15, 441; 1898, Volume 18, 228-229; 1929, Volume 45, 137.

3. Likewise, Josephine Dinner, aged 49, was a match worker with four children aged 19, 18, 15 and 11. All four children were working. The 11-year old girl was a laundress, the most miserable sort of unskilled female labor.

4. The wage structure at the factories was arranged according to the job, not according to seniority. Perhaps those men who supported a wife or a wife and children had moved into the highest paid (and most dangerous) jobs as chemical dippers. In this job category, the wages were eight francs a day, which was the equivalent of the average male and female wage combined.

5. There was no direct evidence of Marie Jouhaux's relationship to a famous match worker named Léon Jouhaux, although the circumstantial evidence is impressive. Léon had a sister who was born in the same year as this Marie Jouhaux and married in the same year as the birth of Marie Jouhaux's first child. We know that the father of Léon worked at the Aubervilliers factory from 1873-1892. In that year, the father fell ill while the husband of Léon's sister died. If Marie Jouhaux were his sister, that could explain why her first three children were named Bouvard, but her fourth child, who was born in 1894, was named Horiette.

If Marie was Léon's sister, she had a disabled father, a mother who worked as a cook, and a teenaged brother. Therefore, she could not have expected financial help from her family. Léon's work history may shed some light on the hiring practices at the Pantin-Aubervilliers factories. We know that there were few workers under the age of 16. Léon began to work at the age of 13, when his father was disabled, but he did not enter the match factory until his sixteenth year, which suggests that the management no longer hired children, even the sons of ex-match worker fathers and widowed match worker sisters. Léon worked for the Aubervilliers factory from 1895 to 1905 when, thanks to his representation of the Trélazé local within the Federation, he became the representative of the *Bourse du Travail* of Angers to the National Committee of the C.G.T. Jean Maitron, *Dictionnaire biographique du mouvement ouvrier francais*, Paris: Les Editions ouvrières, 1975, volume 13, 122.

6. *La Justice*, March 21, 1898.

7. One hundred and eleven match workers (72 women and 49 men) were traced from the 1891 census to the 1896 census. Ninety-five (62 women and 43 men) were still at the match factories. Four women had jobs which may have been at the match factories: one was listed as a supervisor, the three others were listed as daily workers at unspecified locations. Six women had become full-time homemakers at the ages of 27, 31, 32, 37, 44, and 64 with no apparent similarities in their household structures except the presence of a working husband. That is, they did not retire at the birth of a new child or the coming into the labor force of a new household member. There is no evidence from personnel records or newspaper accounts that these six were among the necrosis sufferers. The men who left were aged 23, 26, 32, 40, 46, and 56. They turned to an eclectic set of occupations: one horsegroom, one driller, one deliveryman, one margarine maker, one trucker. The 56-year old retired.

8. Josephine Magar was an unemployed 17-year old in 1891. Catherine Gilligen was a daily worker in 1891; again it must be mentioned that often a name which the census-taker listed as a daily worker was found in the personnel dossiers as a match worker in the same year. Although Gilligen's personnel file was not found, we should keep in mind that she may have simply moved from irregular employment in one of the match factories

in 1891 to a permanent position. Marguerite Hamann was a 14-year old box maker in 1891 and the sister of a match worker. Two sisters in their fifties came in from a tobacco factory. Of the new male match workers, four had been under the age of 14 in 1891; Gaspard Guehol and August Legros had been daily workers. René Grugeard had been only 16 and employed making paper. Jacob Generet had worked as a tripes butcher. Francois Perié was the only man who had enjoyed a high-paying job; he had been a mason. His wife and one daughter, however, had been match workers in 1891. By 1896, he, his wife, two daughters and a step-daughter all worked for the State in that capacity.

9. *Le Petit Bleu* of December 18, 1899 reported a conversation with a pretty 20-year old woman match worker who had been eagerly examining her teeth every day for a year. She was happy to have just discovered the beginning of a necrosis, which she expected would give her the right to a 1,500 franc pension within 12 to 18 months. With this pension she intended to realize her plan to marry the man who had been waiting for just this event since his return from military service. Although the young woman was overly optimistic about the sum she could expect (the highest pension awarded to a man was 950 francs; women received less), this anecdote suggests several interesting attitudes among some of the match workers. First, they were not worried that damage to their teeth and jaws was only the beginning of a long series of ailments resulting in death. Second, some of them planned their marriages around economic considerations. This evidence reinforces the picture of the match workers of the 1890s as the very opposite of the miserable victims of industrial society and its destruction of the family.

10. *La Fronde*, December 24, 1897.

11. H. Desoille, J. Scherrer, R. Truhaut, *Précis de Médecine du Travail*, Paris: Masson et Compagnie, 1975, 320; *Encyclopédie Médico-Chirurgicale*, Volume II, Paris, 3.

12. Of the 13 women match workers who lived with their boyfriends and children in 1891, information was available on only four in the 1896 census. The others were no longer at the same address. Of the four who remained, three had married their boyfriends and were living with the same child as in 1891 but had borne no more children. One, Angèle Vial, was no longer living with her boyfriend named Lacour but with another man named Lejeune.

The child who had been in her household in 1891 was named Roussel. He was no longer with her in 1896 but the fact that he did not have her name or the name of her 1891 boyfriend opens the possibility that he may have been a nephew, a cousin, etc. Of the three male match workers who had children with their live-in girlfriends in 1891, only one was at the same address in 1896. He too had married the mother of his children.

13. Angèle Vial, age 32, was living with a 31-year old boyfriend and a four-year old child. Their case has been described in the previous endnote. Marie Hartz, age 40, lived with Francois Kieffer, age 36, and Marie Kieffer, age seven. Marie senior was classified by the census-taker as "mother" rather than "wife". Therefore, in my study she is listed as a girlfriend, although the fact that the child bore the name of her male companion suggests that the child may have been their legitimate daughter. Victorine Lepape, age 46, was a widow who lived with her legitimate 11-year old. Her boyfriend was a divorcé. Marie Scull, age 47, was living with her 28-year old daughter Maria Renelle and a 43-year old man named Charles Boulanger. Because the census-taker listed them as mother, child, and father, and because they had three different last names, they had to figure in my study as boyfriend and girlfriend although they are hardly the typical pre-nuptual couple with one toddler. Five years later, in the 1896 census, Marie senior and Charles were listed as married. The grown daughter was gone.

14. Jacques Boire, age 41, had children aged three and one. Nicolas Bohn, age 45, had a five-year old, a three-year old, and a one-year old. Marie Erhart, age 42, had two children under the age of four. Marie Georgen was 40 years old with a five-year old son. Marie Jolivet had a baby at the age of 41.

15. *Annuaire statistique*, Volume 15, 1892-1894, 15. Results from the 1891 census: Children of men and women, married 0-5 years including widows, widowers, and divorcés.

Number of children unknown	1.7%
No children	32.0%
One child	38.0%
Two children	19.0%
Three children	6.4%
Four or more children	2.8%

16. There were nine match workers who were unmarried in 1891 but

married according to the 1896 census. Four were still childless: 31-year old Mathias Colman, 31-year old Catherine Kahl, 30-year old Julie Lazar, and 40-year old Marie Moro. Four had one child: Mathias Masson, age 26 had a newborn; so did Paul Rondet, age 31. Marie Dalem, 23 years old, had a five-month old; Anne Goudembourg, age 32 , had married the father of her son, who was six years old in 1896. The only recently married parent of two was Emile Loegel who fathered both children while suffering from an advanced case of necrosis. On July 22, 1893, he married 22-year old Barbara Veiller, a fulltime homemaker. In February 1894, one daughter was born; ten months later he was authorized to stop work because of his necrosis. In November 1895 his second daughter was born; 16 months later he withdrew from the match factories permanently with a necrosis pension of 950 francs. The information on the French birthrate is from the *Annuaire Statistique*, 1892-1894, Volume 15, 15.

17. The numbers were too small to be significant because of the match workers' tendency to change address between censuses. The time available to me for research did not permit a thorough search to connect all of the match workers who were found in the 1891 census with their new address in 1896. Only if they lived within the neighborhoods of the densest concentration of match workers were they found again. Of match workers in 1891 who had a child under the age of one, only two mothers were found in the 1896 census. Of those two, one had successfully raised the infant to age five. There were three children aged twelve to twenty-four months in 1891. All three survived. Of the two children aged twenty-four to thirty-six months, both survived. For comparison, the national mortality rate for children aged 0-12 months was 16-18% between 1891-1896. *Annuaire Statistique*, 1903, 4. Dr. Francois Arnaud produced comparative figures on stillbirths among the offspring of Marseille women match workers and women match box makers from October 1893 to December 1894. For 1895 and 1896 he compared the infant mortality rate among women whose worksite was exposed to white phosphorous and those whose worksite was not exposed. Every year those subjects who were exposed to the poison had a slightly higher birthrate and a lower infant mortality rate. Although he admitted that he had accepted the Administration's figures without verifying the data, it is worthwhile to note that the infant mortality rate among both

sets of women workers was higher than the national average.

	# Women Workers	# Births	Stillbirths	Infant Mortality
October 1893-December 1894				
Match Workers	350	59	10%	unknown
Match Boxmakers	120	17	12%	unknown
1895				
Women exposed to white phosphorous	290	54	unknown	17%
Women not exposed	170	26	unknown	27%
1896				
Women exposed to white phosphorous	263	38	unknown	26%
Women not exposed	191	21	unknown	28%

Dr. Francois Arnaud, *Etudes*, 238.

CHAPTER IV: THE UNION STRUGGLE 1888-1895:
FOUR STRIKES

The victory over white phosphorous poisoning was possible, despite the low priority of occupational health and safety issues in the 1890s, because the match workers benefited from some of the structural conditions which we have discussed: the ability to exert economic pressure on the Ministry of Finances, the prevalence of women of childbearing years in the factories, and the growing public awareness of occupational risk.

But why were the match workers in France able to exert pressure and fight successfully while the match workers in other countries were not? And why was reform effected in the match industry earlier than in those industries which dealt with the other two major toxic chemicals: lead and mercury? The answer lies both in the structural conditions specific to the French match industry in the 1890s and in the correct tactical decisions made by the match workers themselves.

First, match workers in other countries were not government workers. Although the French government was willing to break protracted strikes in key industries in the private sector, for the public sector such as the railroad workers, the merchant marine, and the tobacco workers, there was a tradition of making exceptions. For example, in the government-owned tobacco factories, there had been twelve strikes between 1870 and 1887. On only one occasion was it recorded that force was used: in a six-week strike in 1875 in Toulouse, the factory was occupied by the Army. More often, however, the matter was either settled internally or the workers insisted that their special status required direct negotiations with a representative from the Ministry of Finances. In 1883 and 1887 such negotiations were necessary to end strikes at the Le Havre and Marseille tobacco factories respectively.[1] When the Lyon tobacco workers formed a union in 1887 and the Marseille tobacco workers did likewise in 1888, their factory directors and police prefects examined their statutes for evidence of illegal intentions but found none. The Inspector of Finances and the Minister of Finances came to the same conclusion.[2] The tobacco workers also seemed to believe that because unions had been legal since 1884, any action which was undertaken in the name of their union was legal. For instance, when the woman vice-president of the Dieppe tobacco local was put on suspension for some small mistake (*une faute banale*) in 1901, she declared that her status as a union officer rendered her untouchable.[3]

Many match workers had been tobacco workers or were related to tobacco workers. After 1890 the match workers assumed that they too benefited from these special privileges. In addition they could rely on a few sympathetic deputies in 1890 and more after the elections of 1893.[4] The match workers lobbied those deputies and the Minister of Finances, a practice which was alien to most working class people. In all of these ways, the match workers used their special status as a means of exerting political pressure.

Another point in their favor was the fact that the economic impact of any change in their industry would be small. A ban on white phosphorous would affect only one industry: matches. In comparison, restrictions on the industrial use of lead would have affected sixty trades.[5] Also by the 1890s France had lost its export market in matches to competition from Germany, Sweden, Belgium, and Japan. South America, which had bought French matches in the 1850s, had erected tariff barriers. (See Chart XV: Exports in Numbers of Matches in 1894, 1896, 1897). These economic conditions of the 1890s also created the material possibility for a ban.

Of course that ban might have been realized eventually without the class conscious actions of the match workers. As Marx wrote in his analysis of English factory legislation:

"Factory legislation, that first conscious and methodical reaction of society against the spontaneously developed form of the process of production, is...just as much the necessary product of modern industry as cotton yarn, self-actors, and the electric telegraph."[6]

Society's reaction to white phosphorous may have been inevitable but it was also conscious. To bring the issue to the consciousness of French society as a whole and the French government in particular required movement on the part of the match workers: both direct action strikes and appeals to the bourgeois state through parliamentary lobbying and the press. The match workers were constant in applying pressure by one means, the other, or both. Although the victory was possible because of the particularly fortuitous set of political and economic conditions which the match workers inherited in 1890, without the solidarity of their labor union, without indefatigable union leadership, and without a politically conscious union membership, necrosis would have continued beyond 1898 as it did in the other match manufacturing countries.

CHART XV: EXPORTS IN NUMBERS OF MATCHES:
1894, 1896, 1897

	1894	1896	1897
Europe: Monaco, Gibraltar	51,854,400	71,211,800	72,221,240 (includes Spain)
Africa: Algeria, Tunisia, Morocco, Senegal, Gabon, The Congo, Zanzibar La Réunion	447,187,060	258,120,720	139,645,360 (includes the Ivory Coast and Obock)
Others: The Antilles, Tahiti	16,416,000	1,706,400	-
Subtotal	515,457,000	331,037,000	211,866,000
Navy	2,584,00	1,920,000	1,752,000
Total exports	518,041,460	332,958,920	213,618,600

Direction Générale des Manufactures de l'Etat, *Compte en matières et en deniers de l'exploitation du monopole des allumettes chimiques pour l'année*, Paris: Imprimerie Nationale, 1894, 97; 1897, 97; 1898, 65.

To examine the workers' actions, let us look at four strike years: 1888, 1893, 1894, and 1895. The gains and losses of each taught the match workers how to proceed.

On May 17, 1888, 200 match workers at Aubervilliers struck over the introduction of an inferior grade of wood. It was more difficult to handle, which caused them to work harder in order to earn their usual wages. There are contradictory accounts of the results. According to one account, the strike failed completely. Strikebreakers came in from another factory; the next day the strike was finished on management's terms. According to another account, the strike was a partial success. After four days, eleven strikers were fired but half of the requested wage hike was granted.[7] What is significant is not whether they did or did not secure some alteration of their working conditions but that the strike, according to both accounts, was short and that the other match factories did not follow suit. At least the neighboring factory at Pantin might have been expected to show solidarity because it was close enough to be roused by word of mouth and because many workers had personal and family ties to Pantin workers. On the contrary, if strikebreakers came in, they might well have been from the factory at Pantin. Therefore, the match workers at Aubervilliers learned from this hasty action that they did not have the strength to stage a long strike and could not count on the solidarity of their fellow match workers.

This lack of solidarity was quickly remedied after the Ministry of Finances assumed direct administration. With the help of the Tobacco Workers' Federation, formed in 1890, the match factories also unionized between 1890-1892. The pattern for each local was as follows. First, personal letters were sent to sympathetic individuals from tobacco or match workers. Then union delegates would speak at a public meeting, often with sympathetic local political figures in attendance. The pacific and legal nature of unionism was stressed. The fact that the Administration was not hostile was emphasized. In this way the tobacco workers had organized seven of their factories by 1890; they had completely organized their industry by 1893.[8]

The match workers relied on the fraternal aid of their more experienced comrades. The membership was eager. The first match workers' local formed in Marseille in March, 1890 with statutes copied from the Tobacco Federation. Marseille match workers then traveled to Trélazé to organize their fellows in October, 1890. The Trélazeans already had pro-union sympathies because many were the relatives of the militant union members in the slate mines.[9] Next the match factory at Bègles joined in 1891. Finally, the workers at Pantin-Aubervilliers simultaneously organized their own two locals by December 17, 1892 and the First National Congress, which was held in Paris from December 26-28, 1892. Pantin-Aubervilliers also copied their statutes from the Marseille tobacco workers' union, secured the headquarters of the Tobacco Workers' Federation for their founding congress, and asked the Secretary-General of the Tobacco Federation to be their Acting Secretary-General until they could hold their elections in February, 1893.[10]

Two delegates from each of the five match unions gathered in Paris for the Congress. Copying the Tobacco Workers' structure, the

Federation of Men and Women Match Workers of the State Manufactures (*La Fédération des Ouvriers et Ouvrières des Manufactures des
Allumettes de l'Etat*, hereafter called the Federation), elected its central committee exclusively from the workers at Pantin-Aubervilliers.[11]
Each provincial union was to communicate with its representative but
a cluster of Pantin-Aubervilliers militants could at any time speak for
their two factories or for the whole. Any local had the authority to
strike alone for as many as four days, after which the entire Federation was obliged to strike in solidarity. This structure, coupled with
the central committee's proximity to the Ministry of Finances, was
to be the bane of the Ministry's existence in the coming years.[12]

At the First Congress they also paid a great deal of attention to
medical and hygiene questions. The delegates reported that at each
factory the workers received some free medical and dental care, but
there was no uniformity from one factory to another in frequency
of dental exams, the freedom to use one's own dentist or doctor, or
the percentage of wages which were awarded as sick pay. Rather
than plan negotiations over these details, the delegates simply voted
that the responsibility for necrosis was entirely the Administration's.
Therefore, they estimated that the Administration ought to ban white
phosphorous by July 1, 1893. In the meantime, they considered that
an appropriate level of sick pay was six francs per man or four francs
per woman per day. This was higher than average wages.[13]

When we realize that the Federation was to fight over the white
phosphorous question for the next six years, we wonder whether they
really believed themselves capable of forcing the Administration's
hand in a scant six months. It is entirely possible that they thought
so. First of all, the factories were already producing a small quantity
of red safety matches (13.6% of the total production in 1892).[14]
No matches other than safety matches were permitted in military
establishments; therefore, the match workers thought the government
had the power to extend that order to the whole country and to
enforce it. They did not address themselves to the problem of
contraband. Whether they were aware of the problem or not, it was
not their responsibility. Second, they had enormous faith in the power
of unionism. They had the example of their comrades in the tobacco
factories who had sometimes won changes in the quality of their raw
materials. The strength of the unionism of both the Tobacco and
Match Federations lay in the rank and file women who were rarely
mentioned by name, but who quickly took offense when they did

not like the materials they were working with or the individuals who supervised them. Once the rank and file was angry, the women more often than the men were eager to walk out.[15]

The Federation of Men and Women Match Workers emerged from their 1892 founding Congress spoiling for a fight. Their national organization was firm. Their allies in the larger and more experienced Tobacco Federation were eager to help. Their demands were set. They had a deadline to meet.

Eleven weeks later, an ordinary incident at the Pantin factory set off another strike. It began innocuously enough on March 18, 1893, when a foreman changed workshops due to illness. For unknown reasons, the women in that workshop objected, shouting at him that he ought to stay at home if he were really sick. He singled out two women for punishments of eight-day layoffs. This only further excited their co-workers who insisted that the days of layoff be evenly distributed among them all. At this point, a neighboring workshop ran out of raw materials and stopped work momentarily. Apparently, those workers became involved in the discussion. Both groups walked off the job. Runners were dispatched to the other parts of the Pantin factory and the Aubervilliers factory to urge them to strike.[16] A meeting was held in the courtyard. Four demands were issued: a 15% across-the-board raise, the firing of all the foremen who had been hired under the private company, a ban on white phosphorous, and no more punishments. Obviously, this was more than a spontaneous disagreement with a sick foreman. The workers asked for higher pay although they were already well paid in comparison with other factory operatives. There was no discussion at this time or at any subsequent time during the 1890s of raising women's wages to or towards men's in this industry where 70% of the workers were women. Everyone just wanted more. *"L'Etat est considéré faire mieux;"* the State was expected to be the Model Boss to set standards for the nation at large.[17]

On the firing of foremen from the private company, it is difficult to tell whether the workers disliked all of them or whether they resented certain individuals. What is certain is that both the tobacco and the match workers had a history of flare-ups between foremen and women workers. The vagueness with which the grievances are described in the existing sources leave one wondering to what extent verbal and physical harassment, perhaps sexual harassment, were facts of life. For instance, the Marseille tobacco union was born in 1887 of a dispute over a "severe and uncouth" (*rude*) section foreman who had already been the object of a protest in 1883. The Bordeaux tobacco workers had walked out in 1889 when an engineer was guilty of "bothersome" behavior (*vexations*) and a section foreman was "tactless". The women returned to work only after the engineer had been sent elsewhere and the Director himself had taken charge. In April, 1893, when a delegate from Pantin went to organize the match factory at Saintines, one of the grievances he heard was of a supervisor who "mistreated" (*malmené*) women workers.[18] The demand that they all be fired was close to the demand that all punishments be abolished. The match workers wanted no authorities over them. The match workers had no *modus vivendi* on power issues, no collective bargaining tradition with written contracts specifying work rules and degrees of infractions.[19] They were trying to see if they had the power to fire an entire level of the bosses. This was a strike for nothing less than to define the power of the union.

An Administration spokesman answered on the next day: March 19, 1893. Wages were already high and benefits were unusually generous: pensions after thirty years; contributions to a savings fund (*caisse d'épargne*); workers' compensation including payments for dental care, maternity benefits, doctors' fees, medicine, and something called "hygienic" beverages (either medicinal gargles for all of the workers or milk for necrosis victims). The Administration also paid benefits to workers who served in the Army and made contributions to mutual aid societies. The spokesman promised that white phosphorous would be phased out eventually as consumer preferences permitted. In defense of the Administration's record of progress, he claimed that since 1890 there had been only one new case of necrosis. This case, however, he blamed on the worker's own carelessness without offering any justification for his accusation.

On the questions of the foremen and the punishments, he was unmovable. He claimed that the Ministry of Finances had incurred an obligation to the foremen when it had assumed direct administration and that without punishments there would be no discipline. In other words, the answers to the four demands were: no, all in good time, no and no.[20]

On March 22, 1893, with the workers still refusing to come back, the management of Pantin escalated the fight. According to *Le Temps*, after refusing the union's demands, management then refused even to receive a union delegation. In a further effort to cut off negotiations, next they fired Ernest Deroy, the newly-elected Secretary-General of the Federation.

The next request for a meeting was met with a refusal on the grounds that the delegation included a non-worker, Deroy.[21] There is no record of any specific charge against him. A 1902 study says only:

> "Mister X...Secretary-General of the Federation, made himself known by his violent acts and formal refusals to obey orders."[22]

The Pantin-Aubervilliers leadership now put into play their double strategy: simultaneously appealing to their representatives within the bourgeois state and calling for working class action. They lost no time in calling on the "workers' deputies" to the Chamber: Antide Boyer (Marseille), Jean-Baptiste Dumay (Belleville), Dr. Antoine Ferroul (Aude), Emile Goussot (Pantin-Aubervilliers), and Antoine Jourde (Gironde), who promised to bring the issues to the floor.[23] Simultaneously, they roused the locals at Bègles, Trélazé, and Marseille. Strike funds flowed in from the tobacco workers and from fifteen or more Parisian unions.

By March 26 the match workers exerted pressure one level higher through direct discussions with Minister of Finances Pierre Tirard; it was a negotiating tactic which these workers were often to repeat. On the phosphorous question, they complained:

> "If a worker is sick, they take away his work and all the time that he's out, he earns nothing. To get his job back, he has to pass the (dental) exam again. It's forced unemployment."[24]

This complaint was exaggerated. According to the minutes of the 1892 Congress of the Federation, some of the workers at Trélazé were already receiving half wages as sick pay. At Aubervilliers the figure was sometimes as high as 100%. However, there were also incidences of no sick pay for able-bodied workers who were told simply to stay away from white phosphorous for a period of time immediately before or after a tooth extraction.[25]

At least on this point, Pierre Tirard was able to offer a concession. He promised that sick pay would become generalized to all the factories. In those cases where a necrosis victim was judged to be able to work, he or she would be moved to a less dangerous post within the factory. However, the answer was communicated indirectly, on March 27, by the Minister to the Director of Pantin and the Deputies who, in turn, repeated it to the workers. Apparently, Antide Boyer and Emile Goussot gave the match workers their assurance that the latter would have satisfaction on all of their demands, including the 15% pay raise. This promise was to lead to later feelings of betrayal.[26]

The Minister of Finances agreed that Deroy could return to work, but there was the problem of saving face for those levels of management which had been superceded when the Federation went straight to the top in its negotiations. Precedent from the tobacco factories could have justified a formal apology and expression of respect from Deroy. But previous firings for alleged acts of insubordination had concerned no worker higher than the vice-president of the Lyon tobacco factory's insurance and strike fund versus no manager higher than an engineer or section supervisor. Since the formation of the tobacco union, management had backed down twice by moving objectionable managers to other jobs. In this case, the Director of State Manufactures, Pradines, proposed a reconciliation between Deroy and the Director of the Pantin factory, Descombes. Descombes answered that he would consider rehiring Deroy only if all the other workers returned to the job first. The Federation members recognized a trap. *"Vive Deroy! Vive la grève!"* was the decision of the Pantin-Aubervilliers locals. A group of women at Pantin initiated a petition asserting that Deroy had in no way disturbed their work by his union activities as management alleged. The Bègles and Trélazé locals voted to prolong the strike over the rehiring of the Parisian.[27]

The stalemate was broken when the Director of State Manufactures, Pradines, suddenly announced his retirement. On March 30, the new Director of State Manufactures promised that Deroy could return; that the original punishments were rescinded; and that there would be an average raise of 15%, without, however, specifying which jobs or which factories would receive raises. Descombes, the Director of Pantin-Aubervilliers, accepted Deroy's return. Their relations continued to be hostile, as might have been expected.

All of the five striking factories returned to work on March 30, 1893. Pantin turned the first morning of work into a media event. In addition to the press, 1,000-2,000 people, delegates from other trades, and two brigades of police were present to witness Pantin's return to work. When the shift bell rang, the central doors to the factory were flung back with a bang. Deroy marched in first, then the women, who usually entered through another door, then the rest of the men. Next, a runner was dispatched to the Aubervilliers factory to give the word that Pantin was back at work without incident so that they too might proceed. The other locals were then notified.[28]

The match workers had learned lessons which became part of their collective memory and influenced their future union actions. First, they had succeeded in halting production or slowing it to a trickle. At Trélazé only one load of finished matches had left and the police had to be called in to escort the delivery truck. Second, there had been little strikebreaking. Only four to eight Trélazeans had continued to work. The only violent incident had been more comic than tragic. On March 23, 1893, a male worker who tried to enter the wood-cutting workshop was rushed by 20 women match workers who carried him 600 meters away from the factory, tearing his shirt and insulting him the whole distance while he protested their violation of his legal right to work. He complained to the police, who inquired, but apparently took no action against the assailants. However, by March 27, 1893, the factory director insisted on ten policemen on the premises at all times and police on horseback at opening and closing times. At Bègles the factory director had transferred the few who wanted to work to temporary jobs at the Bordeaux tobacco factory.[29]

Third, they had been able to count on the solidarity of other unions. Some of the tobacco workers at Bègles were the parents of striking match workers. They had resisted their director's threat to pressure their children back to work or lose their own jobs. Finally, the strike had not hurt financially. The contributions of the Trélazé slate miners to the match workers' union reinforced the confidence and labor solidarity of the latter. Only six weeks later, on May 14, 1893, the two unions were able to give financial help to striking spinners. The Marseille City Council put 1000 francs at the Mayor's disposal for needy families of match workers. The Paris City Council was asked for 10,000 francs; the Seine General Council allocated 5,000 francs. All 21 locals of the Tobacco Federation sent contributions to the Match Federation. Each striker received an average of three francs per day in strike funds.[30] The match workers had struck for nine days without financial hardship, loss of jobs, or much disruption of personal relations. Perhaps most encouraging of all, the fledgling Match Federation had brought itself to the attention of the rest of the labor movement. In the flush of victory, the last unorganized match factory, Saintines, with 210 workers, voted to affiliate.

On the specific issue of white phosphorous, they had planted some seeds of victory. First, they had drawn public attention to the issue. They had established a relationship with sympathetic newspapers which emphasized their sensational claims: that they needed a raise because the average woman earned 75 centimes per day (a more realistic average was 3 francs 25 centimes per day), that children who were born to the women in the dipping departments all died before the age of five, and that 35% of necrosis victims died while the remaining 65% were never again able to use their jaws.[31] Second, the Administration's guarantee of sick pay opened the door to negotiations for other damages: early retirement, the support of dependents, and funeral expenses. Finally, by granting industry-wide benefits rather than arguing different conditions at each factory, the Administration had given the Federation reason to stand together in the future.

The Ministry of Finances seemed to have conceded very little: a vague promise on wages, the uniformization of an already granted benefit on sick pay, and amnesty for the strikers; in reality, the Ministry had given the Federation an advantage which it would press in the years to come. In addition, the Ministry had been politically embarassed at a time of growing labor unrest, increasingly frequent strikes and increasing violence on the parts of strikers, the police, and the Army. (See Chart XVI: Strike Results 1890-1912). We must remember that the syndicalist part of the labor movement believed that the only path to the future social republic lay precisely in greater and longer strikes which would culminate in the General Strike. A significant part of the bourgeoisie feared that they were right and saw the government's inability to stop strikes as a sign of weakness. While *Le Temps* ran a front-page editorial reprimanding the government for ushering in socialism with its monopoly, *Le Rappel* attacked from the left by claiming that the firing of Deroy was a breach of the law on the freedom to engage in trade union activities. *La Justice* compared the French Republican boss to Bismarck, the national enemy, who wanted State workers to fall under the same non-union statutes as the military. *Le Socialiste* compared the State unfavorably to the owners of the ironworks at Rive-de-Gier, who had just shown particular brutality during a strike.[32]

Therefore, Pierre Tirard may have acted with such alacrity to show that indeed a republican government could govern; could keep labor peace; could run its own industries efficiently, quietly, and profitably. Perhaps this was why Ernest Deroy was allowed to return to work without apologizing and why the socialist deputies were encouraged to tell the Federation that they had achieved satisfaction on all of their demands.

CHART XVI: STRIKE RESULTS 1890-1912

Year	Total number of strikes	Succeeded	Compromise solution	Failed
1890	313	82(a)	64	161
1891	267	91(a)	67	106
1892	261	56(a)	80	118
1893	634	158	206	270
1894	391	84	129	178
1895	405	100	117	188
1896	476	117	122	237
1897	356	68	122	166
1898	368	75	123	170
1899	739	180	282	277
1900	902	205	360	337
1901	523	114	195	214
1902	512	111	184	217
1903	567	122	222	223
1904	1026	297	394	335
1905	830	184	361	285
1906	1309	278	539	492
1907	1275	263	490	522
1908	1073	185	324	564
1909	1025	217	385	423
1910	1502	307	598	597
1911	1471	261	529	681
1912	1116	193	382	541

(a) Results unknown for some strikes in 1890, for 264 strikes in 1891, for 254 strikes in 1892.
Information from *Annuaire statistique de la France*, Paris: Imprimerie Nationale, 1912, 42*.

In reality, the match workers saw no raises until they had protested again. On April 26, 1893, some of the Trélazeans complained to their director, assuring him that their deputy had confirmed the 15% pay hike. The Director of Trélazé responded that he knew nothing about it. They grumbled about calling another strike but left the decision to the central committee. Deroy and four other men from Pantin visited the new Director-General on May 7.

They were surprised to hear that there had not been a firm promise of a 15% across-the-board raise. After a two-hour meeting with the Director-General at which they passed in review 48 grievances, the latter agreed to raise wages in a few especially unhealthy workshops. Those who filled the presses would have seven percent raises; the dippers, who were the most exposed to hot white phosphorous fumes, were accorded ten percent; the driers were to receive eight percent. The Pantin-Aubervilliers locals announced a party to celebrate.[33]

The next evidence of worker dissatisfaction occurred at Saintines, where the new local held a four-day strike from April 15-19, 1894, in the hopes of replacing their Director. The workers failed to achieve what the older locals had accomplished. They were not able to oust the Director, but they did succeed in testing their organization, then ended the strike before the Federation's statutes would have obliged the five other locals to follow suit. Saintines' delegates to the Second National Congress, which took place the next week in Saint-Denis (April 23-28, 1894), could consider that they too represented an established militant local.

At the Congress, discussion of "the dental question" revealed that there was still no uniformity in medical care and medical payments despite the ministerial promise of March, 1893. At Marseille a dentist visited once a month. Examinations were voluntary, but those workers who refused were denied compensation in case of necrosis. At Pantin-Aubervilliers, a dentist came more frequently but workers were obliged to pay for their own extractions. The delegates from Bègles were especially dissatisfied with their extraction-happy dentist who pulled teeth needlessly at the slightest sign of decay even without signs of necrosis. The delegates passed resolutions in favor of free exams, fillings, extractions, and medicinal gargles. They demanded sick pay equal to daily wages. They again visited the Director-General to present their demands. He again assured them that he was doing his maximum, but that he was constrained by his budget and consumer preferences.

The complaints about extractions reflected disagreements within the medical profession about the relative advisability of fillings over extractions and the necessary period of isolation from white phosphorous before and after either form of dental care. By 1894, a doctor who specialized in dental disease had made himself known to the match workers. An article by Dr. Emile Magitot was published as part of the Collected Resolutions of the 1894 Congress. Thanks to

Dr. Magitot, the Match Federation members learned that the most important medical societies had already declared their preference for a ban on white phosphorous; that more stringent observation of hygiene would, in his opinion, diminish or even eliminate the problem; and that extractions or surgery on affected bone tissue might be counterproductive.[34] This new knowledge may have been the source of the complaint from the Bègles delegates. It was also probably the inspiration for the next incident at Pantin-Aubervilliers on May 18, 1894.

Six hundred of the approximately 740 workers at Pantin-Aubervilliers, the dirtiest factories with the highest number of necrosis cases,[35] refused to let the company dentist, Dr. Noiret, examine them. They claimed that he was a brute who, like the dentist at the Bègles factory, yanked teeth needlessly. Management threatened 22 workers with 40-day layoffs. The workers stood pat, Dr. Noiret resigned, the punishments were withdrawn, and Pantin-Aubervilliers was back at work on May 20, 1894.[36] Once again, despite the recalcitrance of the Director-General to make any concessions, the workers of Pantin-Aubervilliers had managed through organization and solidarity to exert their will over part of their working conditions.

The match workers did not trust their health to their factory directors and factory-appointed doctors and dentists, although conditions had improved since 1890. Some of the factories now had lunchrooms well separated from workrooms; the Federation wanted all of the factories to do so. Some of the factories had dressing rooms; the Federation wanted two sets of dressing rooms, the first for street clothes, the second for work clothes so white phosphorous would never inadvertently be brought home. The Administration provided soap, protective aprons, and beakers of turpentine in work areas;[37] the Federation wanted better quality soap and more frequent refilling of the turpentine beakers. They also wanted a liter of milk per day per worker and paid leave far from the factory in pure country air for necrosis sufferers.[38]

Fewer workers were exposed to white phosphorous than in 1890. The Administration was acting in good faith by trying to influence consumer preferences in favor of red phosphorous matches. Since 1892 they had increased the percentage of red phosphorous matches in production and lowered the price of red phosphorous matches.[39] The factories at Trélazé and Saintines were no longer producing white phosphorous matches in 1894, but these changes were all too gradual to placate the workers. At Trélazé there had been five new necrosis cases and at Saintines there had been one since 1893. Because the decreased production of white phosphorous matches was offset by an increased incidence of necrosis cases, the workers settled in for a period of making the continued production of white phosphorous matches as expensive as possible.

Obviously, the Ministry of Finances had inherited from the private company a problem greater than they had realized. Since cases of necrosis could appear months or even years after a worker's last exposure, there was no means of discerning whether they were paying for cases which had been provoked by the pre-1890 lackadaisical attentions of the private company; if their improved sanitation and medical care were ineffective; or if, as Dr. Magitot began to claim at this time, the extractions themselves were creating the conditions under which necrosis developed. (See Chart XVII: Necrosis Cases Per Year Per Factory 1890-1895). In the meantime, the workers used every concession to the maximum. Payments to doctors rose an average of 17% per year from 1890-1894. Costs for medicinal mouthwashes rose by 18% per year. Increases for "workers at rest for care of the mouth" rose by 285% in 1891, by 246% in 1892, by 385% in 1893, and by 208% in 1894. (See Chart XVIII: Institutions for Personnel Welfare in Francs 1890-1894, Graph II: Cost of Workers' Compensation for Mouth Injuries Compared to Cost of Other Personnel Welfare 1890-1894, Graph III: Ratio Medical Care to Profits 1891-1897). On March 2, 1895, the Federation petitioned the Chamber of Deputies on the delay in the substitution of a non-toxic formula, but received no satisfactory answer. They were already arguing that the Administration would save money by the exclusive use of red phosphorous. Although the cost of red phosphorous was slightly higher than the cost of white phosphorous, by 1894 the Administration was spending more than that difference on sick pay. The concession of sick pay, fortified by the strike of 1893, was being used generously by workers and company doctors, but no relief was in sight.

It would take more than money to force the Administration's hand. It would take another strike, a harder one: a strike that would impress the public but would also demand greater sacrifice from the locals.

Again a provocation arrived in the spring. The Ministry of Finances had indicated to those deputies who questioned the wisdom of a State monopoly in 1889 that the match factories of the French State would use French wood whenever possible. However, Pantin and Aubervilliers were still using Russian wood. It should be remembered that a change in the quality of wood at Aubervilliers in 1888 had occasioned a brief strike. When on March 11, 1895, the Director of Pantin-Aubervilliers again introduced French wood, the workers again insisted that it produced a higher percentage of waste; waste decreased their productivity; lower productivity meant lower wages because they were still paid for piecework. They called a strike.

CHART XVII: NECROSIS CASES PER YEAR PER FACTORY
1890-1895

	1890	1891	1892	1893	1894	1895	Total per factory
Aubervilliers·-	-	1	3	-	-	4
Bègles	-	-	-	-	-	-	-
Marseille	-	-	1	-	1	-	2
Pantin	1	-	1	-	13*	24*	39
Saintines	-	-	-	1	-	-	1
Trélazé	1	-	-	3	2	0	12
(relapses)					4	2	
Yearly total	2	-	3	7	20	26	58

* light cases

Data from Direction Générale des Manufactures de l'Etat, *Compte en matières ...1894*, Paris: Imprimerie Nationale, 1895, 1x.

	1890	1891	1892	1893	1894	1895	Total per factory
Aubervilliers			10				
Pantin			1				

Data from union representative Jacques Aschbacher at 1894 Federation Congress.

	1890	1891	1892	1893	1894	1895	Total per factory
Pantin-Aubervilliers					32	123	

Data from *Le Temps* August 6, 1896. These higher figures may represent workers who received sick pay for dental conditions which were under observation but which did not result in a diagnosis of necrosis.

CHART XVIII: INSTITUTIONS FOR PERSONNEL WELFARE
IN FRANCS 1890-1894

	1890	1891	1892	1893	1894
Retirement and savings fund	35,766	62,894	73,918	78,550	93,982
Maternity benefits	1,160	2,160	2,620	2,900	4,100
Benefits to workers in the military	692	1,163	1,432	1,268	2,763
Gratuities	1,400	2,240	2,585	2,965	4,345
Doctors' fees	4,948	5,909	6,209	6,636	8,258
Medicine	2,715	4,099	4,208	8,684	10,377
Medical beverages	4,595	7,323	5,581	8,353	7,867
Mutual aid societies	-	141	521	346	296
Workers' libraries	-	-	-	-	400
Workers' loan society	-	-	-	23	46
Miscellaneous	-	12	4	192	73
Workers' compensation for mouth injuries	685	1,960	4,813	18,526	38,530
Workers' compensation for other injuries	50	855	600	710	3,605
Totals	52,013	88,755	102,491	130,152	174,643

Data from Direction Générale des Manufactures de l'Etat, *Compte en matières...1894*, 108-9.

At the same time the Chamber was discussing the budget of the Ministry of Finances. Socialist Deputy Marcel Sembat (Seine-et-Oise) took advantage of the discussion to propose a 20% raise and the eight-hour day for workers in all State Manufactures. His amendment failed by 377 to 134 but that was a sizable minority in favor of such advanced demands.[40] Deputy Emile Goussot also proposed an amendment to improve the wages and benefits of the match workers. In the course of his argument, he read into the record excerpts from Dr. Magitot's recent report to the Academy of Medicine. First of all, Dr. Magitot believed that there was a pathological, although not disabling, condition called phosphorism and that all of the match workers had it; therefore, any lesion to the bony tissue of the mouth could develop into a necrosis, regardless of the other sanitary precautions in force. The only antidote was the elimination of phosphorous from the worker's body by removing the worker from the premises and giving him or her plenty of fresh air and milk. To further stress the need for an increase in the budget for the State Manufactures, Goussot raised the issues of maternity and the national defense. If necrosis were eliminated:

"...one (would) no longer see these women with livid complexions, worn out by illness, who will one day give birth to new generations carrying the deadly germ, already infected not only physically but intellectually. The phosphorous fumes circulate in the lobes of the brain and that will one day produce anemic generations of paper soldiers, of incapable workers for our agriculture and industry...What I ask of the Chamber is not a partisan proposal of a special interest group: it is a humanitarian act."[41]

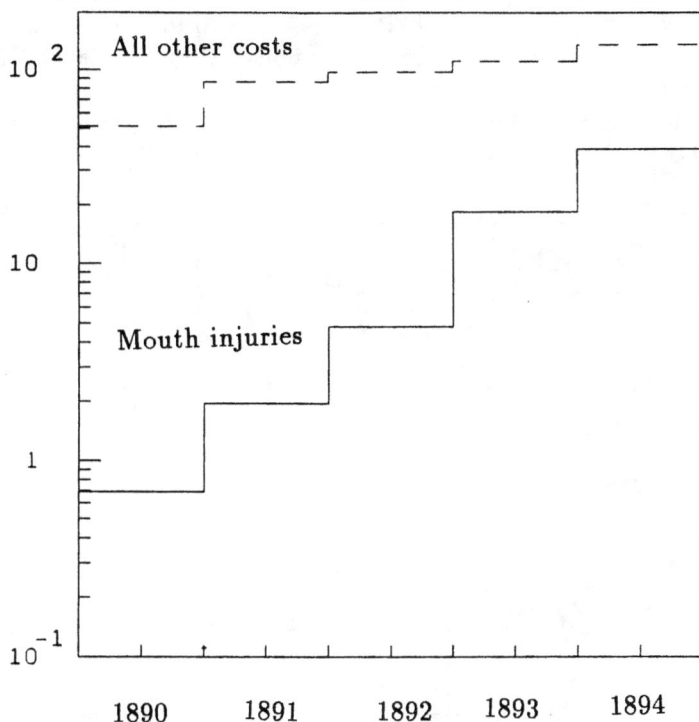

GRAPH II: Workers' Compensation for Mouth Injuries Compared to Other Costs for Personnel Welfare 1890-1894 in thousands of francs

This argument, in the heat of the strike, impressed the Chamber sufficiently to secure a 50,000 franc special allocation for a research committee to examine alternate matches. Thus, only in the third consecutive strike year did the government start to search seriously for solutions.

From March 11-14 Pantin-Aubervilliers struck alone. A worker-management team examined payslips. Three newspapers reported three decisions: that wages had dropped, (*L'Intransigeant*, March 16, 1895), had risen but not as much as expected at the end of the 1893 strike (*L'Echo de Paris*, March 24, 1895), had dropped but not as much as the Federation claimed (*La Justice*, March 18, 1895). The central committee again declared a strike on March 18, claiming that at any rate the use of French wood caused overwork. Although the quality of wood had been a source of discontent before, perhaps the wood was a pretext. The Pantin-Aubervilliers leadership felt that the time was right to gain concessions: they also asked for 5 francs 50 centimes per day (presumably for men) and a ban on white phosphorous. All of the locals went on strike except Trélazé and Beĝles who fell into step at the end of March.

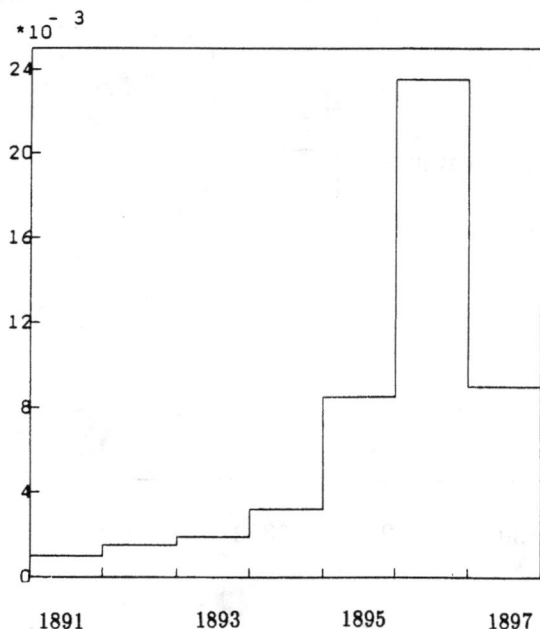

GRAPH III: Ratio of Payments for Medical Care to Profits 1891-1897

There was a festival atmosphere in the northeast working class neighborhoods of Paris where the match workers and their sympathizers lived. At daily meetings, sometimes in the Paris *Bourse du Travail*, reports from the central committee were read and socialists and anarchists made speeches.[42] On April 3, 1895, the Labor Committee at City Hall accepted a proposal that it raise 5000 francs to help necrosis victims. Funds flowed in from other unions and were triumphantly announced. The tobacco workers were again especially generous. The Limoges tobacco workers sent 50 francs and promised another 50 every payday until the bitter end. One tobacco worker sold her wedding ring for 93 francs so that a match worker could pay her rent.[43] An eighteenth *arrondisement* soup kitchen invited strikers and their families to eat for free. Pantin-Aubervilliers shopkeepers gave credit. La Villette butchers donated meat. On April 10, 1895 there were 1600 francs in strike funds with more coming in every day. On April 14 there were 4000 francs. Enthusiastic telegrams from the provinces were read at meetings. Minister Maurice Ribot was backed into a corner where he would have to make concessions or resign; this strike of the proleteriat versus the State foreshadowed the ultimate struggle; let the government lose a million francs but the match workers would never capitulate. It was reported that matches were in short supply in Paris on March 21, in Trélazé and Argenteuil on April third, in Bègles on April 14.

On at least one occasion, feminist journalists spoke and brought 45 francs from a small feminist organization, *La Solidarité des Femmes*.[44] It was, after all, a strike in a majority-female trade and among the worst necrosis sufferers were several women. On April 4, the central committee presented four necrosis victims to the deputies from the Seine and the press. This event was their most successful seizure of public attention to date. It touched that part of the public who usually considered State workers as *budgetivores* or greedy feeders at the public trough. As *l'Intransigeant* described the scene: .

> "In particular, the appearance of the women was horrible: one of them, who had recently undergone an operation, still had her bloody jaws covered with iodine-soaked cotton; another no longer had a nose".[45]

In its next issue, *L'Intransigeant* ran a page one fictional account of the life of a woman match worker under the title "Necrosis". Because of its language similar to the sensational *romans feuilletons* or serialized romances which readers were accustomed to finding on the front page and because of its clear accusation that the State was "assassinating hundreds of women and children" exactly as it did to the heroine of the story, the article is worth quoting at length. It began with a beautiful and virtuous young girl whose family was overjoyed to place her in a match factory.

"She is vigorous...her teeth...are like a row of pearls...but soon, after only a few weeks, she feels headaches, one tooth gives her pain... she grows accustomed. After the first tooth, it's a second, the gum is attacked, the lips become bloodless, the bones are pierced...Five years have gone by. What has become of the beautiful girl with the lovely eyes and teeth like pearls? A miserable creature with stinking breath, a deformed mouth, bleary eyes, a purplish nose..."[46]

The message was clear: the flower of French womanhood was being struck down by the very government which claimed to represent the people.

The successful publicity campaign and jubilant meetings of the Pantin-Aubervilliers unions overshadowed the fact that the Federation was not meeting with success in its negotiations. More Belgian matches were brought in by the government: 50 cases on April 3, 1895; 2000 cases on April 7.[47] Even 4000 francs in strike funds came to only two francs per match worker or seven francs per Parisian match worker. Efforts to lobby the Minister of Finances directly met with his refusals. Deputy Gustave Mesureur, president of the *commission du travail*, referred the Federation back to the Ministry of Finances. Also Deputy Emile Goussot, their usual go-between with the Minister of Finances, was making no progress in getting a 150,000 francs special allocation for wages. The small (122 workers) new factory at Aix-en-Provence, which opened in 1894, was not yet unionized and was not on strike. Some of their matches were sent to Bordeaux where they undercut the Bègles strike. An April 1 effort to meet with Félix Faure, the President of the Republic,

"...we urge you to accept, Mister President, the respectful regards of workers who are devoted to the Republic."[48]

met with a polite refusal. On April 10, Félix Faure declined to serve as arbitor. Although he carefully affirmed their rights as citizens to his attention on other matters, he considered that their status as workers for the Ministry of Finances required them to express their work-related demands through that channel. Finally, on April 29, Emile Goussot, Antide Boyer, and Maximilien Carnaud again spoke to Alexander Ribot on their behalf. He refused to make any concessions.

By the end of April, the Federation could not sustain the public enthusiasm which was so necessary to a continuing supply of strike funds. The left public turned its attention and financial support to other strikers such as the Parisian omnibus workers who walked out on April 26. The only sign of progress was that the government-appointed research committee held its first meeting on April 25. The central committee publicized this as if the researchers had already banned white phosphorous: "the ban on white phosphorous is a great peaceful victory for us..."[49] Emile Goussot told the Federation leadership that the Minister of Finances was in reality sympathetic to their wage demands but had to preserve his authority by awarding a wage hike only **after** they called an end to the strike. Believing him, the central committee reported:

> "...wage adjustments will be examined sympathetically after our return to work. We have the word of our elected officials who will, if necessary, remind the President of the Council (of Ministers), who has shown himself favorable to our demands."[50]

The Pantin-Aubervilliers membership, voting both for their own two locals and for the Federation, seized on this promise and voted to return to work on May 2, 1895. They telegrammed the other locals that they had won and authorized their return to work.

The atmosphere in the provinces was less enthusiastic. Although the national newspapers covered the Parisian strike and accepted the central committee's press releases, the departmental archives of the Maine-et-Loire[51] reveal that from the point of view of Trélazé, Pantin-Aubervilliers had called their strike without sufficient preparation, had chosen a demand of interest only to its own workers, and had communicated poorly with its locals during the strike. The leadership had been so out of touch that only the membership's sense of solidarity and some timely personal exhortations from the leadership saved the appearance of a militant five-week strike. We know that the other provincial locals had difficulties similar to those at Trélazé because members of the central committee had to rush to Saintines and Marseille as well as Trélazé in order to rally support.

The first problem was that the leadership at Pantin-Aubervilliers had ignored Trélazé's lack of dissatisfaction in February, 1895. On February 25, we find the prefect of the Maine-et-Loire in possession of a copy of a letter from the Federation leadership in which they reproached the Trélazé leadership for not communicating. In addition, he saw letters which were addressed to Trélazé militants Louis (Jean-Pierre) Bridier and Madame Ménard[52] urging them to send their grievances to Paul Rondet,[53] their representative on the central committee, in preparation for a meeting with the *Direction-Général des Manufactures*. However, the archives indicate no answer from Trélazé. Indeed, the "special officer" or police spy, (*commissaire spécial*) reported that the workers at Trélazé were content because the "paternal administration of Mr. Sévène has disarmed many hostilities..."[54]

The second problem was that the issue which had set off the anger of the Pantin-Aubervilliers workers was the introduction of French wood. This was of no interest to the workers of Saintines and Trélazé, which were located in wooded areas, had sawmills attached, and were concerned about the jobs of their sawyers.

On March 19, 1895, when the strike was again declared, the police superintendent at Trélazé reported to the Minister of the Interior by coded telegram that the Trélazé militants Léon Halopé and Aubert (no first name given)[55] were traveling to Saintines with the intention of arguing against the abandonment of French wood. This was in violation of the union statutes. First, the local at Trélazé was obligated to strike in solidarity. Second, all communication should have been centralized through their representative on the central committee. On March 26, with their two delegates still in transit, the rest of Trélazé's leadership sent telegrams to Bègles, Marseille, and Saintines asking if they were striking over the wages in Paris. Bègles wired back immediately that their strike was unanimous, although on the same day a letter from the Bègles local to the central committee revealed that they had not been able to call their members out.

In the meantime, also on March 26, a police spy filed a detailed report on an ugly fight at union headquarters in Pantin. Halopé and Aubert arrived at their meeting with the central committee two hours late. They had been with their factory director, Sévène, "convincing" the *Direction Générale des Manufactures de l'Etat* of the superiority of French wood. When they arrived at Pantin, the central committee members Ernest Deroy and Jacques Aschbacher reproached them their breach of the statutes and threatened them with sanctions at the next Federal Congress. Trélazé's representative on the central committee had not been informed that Halopé and Aubert were holding a private conference with Saintines. Had they been in touch with their man in Paris, Deroy and Aschbacher continued, they would have known very well that the demand on wood was only a smokescreen for a wage hike. The wage hike would have been already won if not for the "traitors" of Trélazé, said Deroy and Aschbacher. A delegate from Saintines, who had accompanied Halopé and Aubert to support them, remained silent.

Halopé and Aubert responded by complaining of Deroy's excessively centralized and secretive politics. They demanded eight days' notice before any future strike. Deroy snapped that strikes don't come off like that. Deroy was right if he was saying that secrecy was important to a strike. However, as the police had copies of every telegram to and from Trélazé, had an informer at every union general assembly, and had even infiltrated this private leadership meeting, the police knew everything anyway. The Trélazeans next complained that since their director, Sévène, had treated them so well, it was difficult for the membership to turn on him. The Parisians reminded them that they were part of a national industry. Sévène would soon be transferred to Pantin; someone from the tobacco factories would be their new director. (This was correct although one wonders how the Parisians knew.) Deroy then called Halopé a drunkard and an anarchist in front of all the members of the central committee and the delegate from Saintines. He read a letter from Bègles in which the Trélazeans were called "quitters" (lâcheurs).

We can understand why the central committee of Pantin-Aubervilliers would treat the representatives from Trélazé harshly, but why did this meeting turn into a bitter personal fight between Halopé and Deroy? Part of the answer lay in the men's political positions. Deroy was trying to make use of sympathetic deputies, several of whom represented the Seine; Halopé was referred to in the Maine-et-Loire departmental archives as an anarchist who constantly quoted Elisée Reclus and Jean Grave. The town of Trélazé was an anarchist stronghold with a collective memory of the 1855 march of the slate miners on Angers to declare the democratic and social republic. Both the match workers and the slate miners of Trélazé spurned the Angers Bourse du Travail until 1903 because the majority of the leadership was reformist. In fact, when in 1892 the majority of the Bourse du Travail had been convinced by the socialist deputies Aimé Lavy and Jean-Baptiste Dumay to sign the manifesto of the Socialist Congress, two dissenting members published a protest against all collaboration with public authorities.[56] In contrast, the Federation leadership took pains to distance their membership from the anarchist agitation of the years 1893-1895. At several points during the 1895 strike, anarchist posters appeared in one town or another: the Federation leadership was always quick to reject responsibility for such agitation, calling the posters provocations which might have been planted by the police.

This anti-parliamentary tradition of the Trélazeans in general and Halopé in particular may explain the virulence of Deroy and Aschbacher's attack on their intelligence, solidarity, and personal integrity. It may also explain why communications from the central committee had been addressed to Louis Bridier and Madame Ménard.

Deroy and Aschbacher were obviously magnetic personalities; excellent public speakers; scrappy, arrogant, energetic men. They were both known for getting into shouting matches with management at their plants. Of the six to ten members of the central committee who were listed in the police archives of the Seine, it was always Deroy or Aschbacher who stood out as the spokesman. Deroy must have been very strong, or very devoted, or both. At this time he was suffering from necrosis, had lost twelve teeth and been on sickleave for six months.[57] Yet he was to travel to Saintines, Trélazé, and Marseille to deliver all-day harangues within the next month. Nothing is known about their abilities to engage in a more collectivist style of decision-making except that they had successfully worked with the rest of the central committee for two years. Clearly, in this incident, they were not interested in treating Halopé and Aubert as members of the collective, but in discrediting them. They succeeded in silencing and humiliating them, but not in convincing them.

Vengeful towards Deroy, the Trélazeans went home grumbling that the strike was nothing but an invention by Deroy, *"monteur de coups"*. On March 29, 1895, the eleventh day of the strike, they convinced their local not to strike unless the central committee replaced the grievance on wood with more general demands and telegrammed this decision to Pantin-Aubervilliers.[58] The next morning Trélazé received a telegram from Saintines informing them that Saintines had voted to strike, 113 to 44, because Pantin-Aubervilliers was willing to accept French wood. But if the strike was not over French versus Russian wood, what were the demands, Trélazé asked the central committee? At three p.m. on March 30, Pantin-Aubervilliers sent this telgraphic reply:

"General demands, no longer question of replacing French wood."[59]

Bègles too wired:

"Stopped this morning. Have you done same? Answer by telegram."[60]

Trélazé decided to strike. Then at five p.m. they received a letter from Saintines offering to continue to work if Trélazé did likewise. They hesitated. The police report on the decision mentioned specifically that Halopé could not forgive Deroy for mistreating him in Paris, particularly for calling him a drunkard and an anarchist.

On April 1, the woman president of the Trélazé local[61] finally sent the obligatory strike notice to the prefect. Director Sévène weakened their resolve by reading at a general assembly a telegram from the Director of Saintines who said that there was no strike at his plant.

What ensued was a story of masterly union leadership to recuperate a badly begun strike and fragile union solidarity. On April 2, 1895, Trélazé was already considering returning to work according to a 7 a.m. telegram from the special police officer to the prefect. At 8 a.m. a letter arrived from Pantin attacking the actions of Trélazé's delegates, Halopé and Aubert. The other Trélazé local officers, now worried about the 300 francs they had spent sending their men to Saintines and Paris, telegrammed Pantin for details. At seven p.m. they received this response:

"Deplorable action concering French wood. They visited engineer (Inspector-General) Bardot. Bad effect. Comrades, unity. Let us continue the strike with strength".[62]

The next morning at 5 a.m., Ernest Deroy arrived. First he met with the Trélazé leadership including the offended Halopé and Aubert. Unfortunately for history, the archives contain no record of that conversation, but by the time of the morning general assembly, the special officer had to report that Deroy had

"soothed the injured vanities while firmly maintaining the principles of federation".[63]

The theme of his talk before the membership was the Federation: the advantage of the Federation over individual unions, Trélazé as the sole breach in the solidarity of the Federation, the freedom of Trélazé to break with the Federation, and his personal subordination to the welfare of the Federation. First he saved face for Halopé and Aubert by assuring the membership that he believed in their good but mistaken intentions in their visit to Paris; the Director of the match factories had cleverly set a trap which they had fallen into. Their failure to show faith in the central committee, which would have supplied conclusive explanations, was responsible for their error. Halopé then performed his self-criticism. He humbly pleaded extenuating circumstances, on which he did not elaborate, and admitted that he had been wrong. This was a far cry from his arguments in Paris and his bitterness only a few days before. Obviously, an arrangement had been made at that early morning closed door meeting so that the Trélazé leadership and Paris leadership could present a united front. After this day, Halopé's name no longer appeared in union communications. During the midday break Deroy telegrammed the central committee that "all is well and Trélazé will hold steady."[64]

At a four-hour afternoon meeting, Deroy again drew the Federation around him like a cape. He told of his trip to Saintines, which was, he assured them, on strike, because the issues no longer threatened their wood. He read triumphant letters from Bègles and Marseille. Reminding the Trélazé general assembly that they were free to decide otherwise, he made clear that he meant free to go alone, adding with pathos:

"leaving their comrades from Pantin (-Aubervilliers) and Bègles to get along as well as they could." [65]

He was referring to the membership's memories of their brief strikes in 1890, 1891, and 1892 before they had joined the Federation. According to Charles Mannheim's study, the results of those strikes had been inconclusive.[66] Deroy finished by assuring them that he had no personal ambitions. In fact, he said that his necrosis was so advanced that he intended to retire from the match factory; therefore, he was retiring from the union.[67] In the name of the general good he called for a vote on the continuation of the strike. The response was unanimous: "men and women workers in admiration at his gift of speech."[68] Trélazé became a strong link in the Federation's chain.

Fifteen days later, on April 18, 1895, there had been no progress in negotiations, as we know. Donations from the Angers *Bourse du travail*, the slate miners, and even 150 francs from the paternal Director Sévène were running low. Pantin-Aubervilliers offered to send Trélazé's representative from the central committee, Paul Rondet. Instead, Trélazé sent another two delegates to Paris, a man named Villarmet (no first name given) and a woman named

Henriette Bernard, to report on the state of the strike themselves.[69] On the 20th, Pantin-Aubervilliers sent 200 francs in strike funds and promised more soon. On April 24 the central committee managed to send 350 kilos of meat or one kilo per Trélazé worker, and another 200 francs. By this time, three weeks after Deroy's visit and the effective beginning of Trélazé's strike, demoralization had definitely set in. Director Sévène had changed from the kindly donator of funds for the families of strikers and time off for union meetings to the strict factory director who cut off their credit at the food cooperative and refused to guarantee any jobs after the strike. The special police officer noted on April 18 and 26 that only the women were willing to continue the strike, claiming that they received more financial support.

However, on April 27, the special officer reported that the membership had decided that they could no longer resist; they would work on Monday, April 29. The meeting was a sad one. As the meat and money which the central committee had sent three days earlier were distributed at this meeting, *"la femme Bernard"* suggested that it was only decent to return them.

"Many of the women felt a certain awkwardness at accepting the help of the central committee just at the time when they were deserting the cause of the Federation".[70]

Although they did not send back the strike support and did decide to work on the 29th, according to a police officer, the membership felt so down-hearted that the slightest hope would keep them out on strike.

Director Sévène suggested that they simply return to work without informing the central committee who might send their central committee representative or, worse yet, the second-in-command, Jacques Aschbacher. Sévène had not understood that he had lost his authority. Trélazé, the straggler to join the strike, the near-destroyer of the Federation's solidarity, had regained its self-respect by renewing with its radical past.

Once again unity was saved by the Pantin-Aubervilliers leadership just when the fight seemed lost. Money, Deputy Albert Walter (St. Denis), Jacques Aschbacher from the central committee, and a delegate from the Bègles local arrived on April 28. (Ernest Deroy was rallying Marseille on the same day, having just prevented the Saintines local from attending a meeting called by their mayor to vote on continuation). At a three and one-half hour meeting Aschbacher proved himself as persuasive a speaker as Deroy. From the police report, there emerges a picture of a more difficult, more tumultuous meeting than the one 25 days earlier when Deroy had turned the local around between his descent from the train and the first few minutes of the morning general assembly.[71]

With Madame Ménard presiding, the meeting began at three p.m. She was not able to call the membership to order; too many were hostile to the speakers and especially to Deputy Albert Walter. The women match workers (220) outshouted the speakers on the podium and on the floor at several points; the police informant noted how noisy and insolent (*bruyantes et audacieuses*) these working class women could be at a meeting of their union on their territory. In a moment of calm Aschbacher managed to shout: "Are you willing to hear me out?" The police informant reported that the men cried no but their few voices were drowned out by the women's. Aschbacher asked whether they would receive their comrade and Deputy Walter. There was a generally negative movement in the crowd. Aschbacher then hammered out his credentials:

"I have a right to be part of this meeting. I belong to the match workers' trade. The match workers are on strike. I am a striker. I have interests to defend. You owe it to me to listen in silence."[72]

Having caught the attention of the general assembly, he suggested that Citizen Walter's presence be put to a vote. At the mention of Walter's name, disorder broke out again. An unidentified voice asked over the tumult what right Walter had to be there, but, recorded the police informant, it was clear that the women wanted to hear Walter. Aschbacher shamed the local:

"Five thousand busdrivers in Paris were able to talk peacefully, and you, who are only a handful, cannot."

He compared them unfavorably to the match workers at Saintines, who were little aware of social questions but had resisted the sousprefect's efforts at reconciliation. He divided the "good" from the "bad" members of the local.

"...Halopé accepted the statutes of the Federation and today he rejects them. You have decided to return to work. You are the only such members of the Federation. I don't accuse the workers who have shown so much courage. The fault lies with your delegates."

Aschbacher reminded them of their own militant history, flattering them about their allegedly advanced socialist education. Specifically, he mentioned that after the strike of 1893 they had not returned to work unconditionally. He held out the promise of a speedy resolution:

"...there are no more matches at Bègles, at Pantin, at Marseille and you yourselves must know how much is left in your storerooms."

Despite this flow of arguments there was one more burst of anonymous disagreement. Another voice tried to fight over parliamentary procedure:

"You have no right to make a speech!...It is a private meeting. You have no right to keep the floor...We want no deputies or journalists (?) among us".

Ménard, the woman president, made no effort to bring the meeting to order. Aschbacher took the floor again, asking for a voice vote on Albert Walter's presence. There was more tumult. Aschbacher did not make the mistake of asking any more questions. He kept the floor, spouted breathlessly for the rest of the meeting, jumped from theme to theme and did not pause until the assembly was chastened. On their own militant history, he said:

"You claim you have never had suspensions, harassment..... When you threw out a foreman named Blanchot, weren't you happy to call on the Federation?"

In particular he challenged the worth of their friendly relationship with their managers and threw in anti-Semitism (the factory inspector's name was Cahen):

"...Your director has told you he makes no guarantees and yet you have dealt with him...You refuse to listen to Walter but you listened to the Director, the boss, the prison-warden...You will get a Jewish director, a Jewish engineer and a Jewish inspector..."

He claimed that the responsibility for failed negotiations between Deputies Maximilien Carnaud, Antide Boyer and Minister Maurice Ribot lay with their Director Sévène, who had telephoned Paris to say that Trélazé was back at work.

To further discredit the "renegade" Halopé, Aschbacher linked him to the refusal of their deputies to attend their meetings and referred to his and Aubert's alleged good time in Paris on the local's money.[73]

"You-delegates-you are guilty because you drank toasts (*des vins d'honneur*) with the Director and by keeping such company gave the impression to the comrades that you had sold out."

He finished on a note of solidarity: theirs, the other locals' and the solidarity of the labor movement in general. He reminded them of the 200 francs and 300 kilos of meat which their brothers and sisters of the Federation had sent. He claimed that the other locals were holding firm; door-to-door exhortations by the police at Bègles had failed; four scabs at Marseille had abided by a vote of 550 strikers and resumed their strike. He claimed that the recent failure of the bus strike in Paris, prelude to the General Strike, was merely due to a few renegades who lost courage.

After this flow of uninterrupted exhortation, Aschbacher was finally able to give the floor to Deputy Walter without opposition. In the assembly, hostility had turned against the waverers: at this moment a group who had walked out of the room earlier returned; they were hissed by the assembly. Apparently, Aschbacher had already delivered Walter's speech because Walter simply reiterated the themes and listed the demands: a ban on white phosphorous within one month, a raise, and the rehiring of all strikers without reprisals.

Still the local was obviously not feeling hopeful so much as chastened. Aschbacher terminated the meeting by offering more strike funds and calling another meeting for the next morning "so we can see who the renegades are". He offered his own services to deliver to Sévène the message that the members would not work the next day. This meant that the Trélazeans really were not convinced so he could not trust them. Indeed, at this moment, according to the police record, the woman president Ménard finally spoke up to say that they were no longer on strike. Albert Walter offered this compromise: they were not to call it a strike but a suspension of labor for two or three days, not peace but an armistice.

On this sorry note, the assembly agreed that they would tell Director Sévène that they would be back at work in a few days but were waiting for unspecified information from Paris. Fifty-one workers who had decided to work voted 29 to 22 in favor of staying out. Sévène went to Angers to sit out the remainder. Aschbacher stayed until the 30th addressing morning and evening meetings.

As we know, on May 2 all the locals returned to work,

"...according to promises made and measures already under-
way, the ban on white phosphorous being only a matter of
time..."[74]

The central committee had run a masterful strike for 43 days. Union
solidarity was greater than ever. Morale was preserved. Trust in local
factory directors was shaken. Solidarity with local *bourses du travail*
and other unions was reaffirmed. The only problem was that there
had been no "promises made". The Pantin-Aubervilliers leadership
had not informed the locals that they had lost the strike.

It was a changed local that Sévène dealt with on May 2:

"...when I consider the significant difference in the attitude
of the personnel towards me these last few weeks and since
the arrival of the worker Asbacher (sic), I am sad to admit
that the results of my two year effort have been destroyed
in a few days by the dishonest and poorly understood
speeches of the delegate from the Federation...The treacherous
insinuations of Asbacher have closed to me the possibility of
benevolence. I intend to reconquer it by the reestablishment
of my uncontested authority at Trélazé."[75]

The Trélazeans announced that they were returning under the
conditions which had been successfully negotiated by their Pantin-
Aubervilliers comrades: the ban, a 10% raise, and no reprisals.
They would restrict their own demands to a few changes in local
working conditions. For example, they wanted a certain Hervé, who
was employed in the dipping workshop, to change places with a
certain Bernard from Pantin. Sévène surprised them by reading a
communication from the Minister of Finances to the Directors of
all the factories in which he wrote that Pantin-Aubervilliers had
returned to work unconditionally. Bridier, the new leader of the
Trélazé local since the presumed resignation of Halopé, immediately
called another general assembly, after which he announced to Sévène
that the membership would not hear of it, and sent a delegate to Paris
to find out what was going on.[76]

On May 3 at 7:30 a.m. the local at Trélazé received this telegram
from the central committee:

"Saw Directors. Return to work. We are arranging all with
Direction-Générale".[77]

After only a few hours of work, hostility broke out again. The worker
named Bernard, whom Sévène called Aschbacher's lieutenant in a
confidential letter to the Director-General, had refused to accept a job

as a general laborer at 3 francs 60 per day. He insisted on a place as a dipper at 4 francs 50. The Trélazeans, unwilling to strike in March, hesitant to continue in April, now struck alone. They considered the job assignment of Bernard as an act of vengeance against them all. These workers who had been hesitant a few days earlier to call a strike by its name now flung at Sévène the threat that their comrades in the fire department might not assure night service. In the match industry that was quite a threat.

Sévène locked up the factory again and went to the town of Angers. On May 7 the Trélazé delegate to Paris came home. He and the central committee had seen documents in the office of the *Direction Générale* which proved that on January 10, long before there had been talk of a strike, long before there could have been talk within the central committee of a need to send a "lieutenant" to Trélazé, Director Brandeis of Pantin had asked Director Sévène of Trélazé to accept the worker Bernard. On March 8 Sévène had written that he was ready to switch the men but Bernard would have to start as a general laborer. On March 23 Bernard had accepted and would have begun on April 4 but for the strike. The worker Hervé, on the other hand, was guaranteed a place at four francs fifty centimes per day. The Trélazé delegate to Paris reproached Bernard for making fools of the Trélazeans. Bernard became a general laborer and was not mentioned again. Everyone returned to work at noon on May 7, but Trélazé had shown its willingness to strike for an extra five days over union solidarity.

Something similar had happened at the Bègles factory. According to their delegate to the central committee, they too had trouble getting their members out on strike at first. According to *L'Echo de Paris* of March 31, Bègles was not on strike. They made up for lost time in enthusiastic public pronouncements throughout April. But on April 30 their workers too were trickling back to the job. Although they followed the central committee's lead in returning at the beginning of May, on May 6 they struck alone for at least three days. *L'Intransigeant* reported they wanted an end to body searches for stolen materials and the rehiring of a woman worker who had been fired for wearing a red ribbon as a sign of her politics.[78]

Although they had been slow to heat up over Pantin-Aubervilliers' ill-chosen and ill-communicated strike, the match workers in the provinces felt their strength in unity. Going back to work unappeased, Bègles like Trélazé had pent up anger which then flamed at the slightest spark. The strikers' four to six weeks of sacrifices seem particularly poignant when we examine the strike report of the Prefect of the Maine-et-Loire.

Yet the prefect was wrong to report "consequences: none." The Match Workers' Federation had hurt profits so badly that 1895 was a year of loss (See Chart XIX: Profits 1890-1902). They had already had the best possible healthcare, if we take into account the medical profession's confusion on the subject: this strike gave publicity to Dr. Magitot's efforts on their behalf. They had gained public attention and sympathy. A government-financed committee of chemists was searching for an alternative match. In reality they had won. We know this from Director-General Favalelli's letter to the Minister of Finances on November 25, 1895. In this letter the Director-General explained that the concessions won by the strikes had driven his production costs so high that he wanted to produce nothing but red safety matches. He would insist on the change even if chemists did not find a more inflammable substitute.[80]

Questionnaire from the Prefect of the Maine-et-Loire to the Minister of Commerce and Industry, Labor Office[79]

Number of workers:	120 men, 220 women All for the first, second, third and fourth quarters of the strike March 30 - May 7
Returned to work:	All
Refused to return:	None
Fired:	None
Cause:	Solidarity with Pantin in accordance with the statutes of the confederation
Demands at start:	1 Ban on white phosphorous 2 Raise in pay 3 Change in work rules
Offer of employer at start:	Return to work unconditionally
Conditions at end:	None
Wages before strike:	Men 5 francs 30 Women 3 francs 10
Wages after strike:	Same
Hours before strike:	10
Hours after strike:	10
Means of bargaining:	Direct negotiations employer to workers? No Direct negotiations employer to unions? Yes Arbitration? No
Source of strike funds:	Donations, gifts, public requests, other unions 2000 francs
Did workers work elsewhere during strike?	No
Consequences:	None

CHART XIX: PROFITS 1890-1902

Year	Profits in francs	Increase in francs	Percentage	Decrease in francs	Percentage
1890	2,324,562				
1891	19,800,690	17,476,128	852%		
1892	20,072,456	271,766	1%		
1893	20,430,752	358,246	2%		
1894	20,582,334	151,582	1%		
1895	20,115,933			466,401	2%
1896	21,131,835	1,015,902	5%		
1897	21,602,076	470,241	2%		
1898	22,426,597	824,521	4%		
1899	23,037,739	611,142	3%		
1900	23,799,898	762,159	3%		
1901	23,713,247			86,651	0.4%
1902	24,558,301	845,054	4%		

Direction Générale des Manufactures de l'Etat, *Compte en Matières... 1897*, Paris: Imprimerie Nationale, 1898, 20 and Claude Réal and H. Rullière, *Le Tabac et les allumettes*, Paris: Doin, 1925, 353.

Even so, to shorten the wait for the discovery and the introduction of the new match, the union would have to keep up pressure. Having to return to work with neither a wage hike, nor a guarantee on the quality of wood nor a guarantee on white phosphorous, but with their members impoverished, the union was not about to call any more protracted strikes. Without the strike weapon they had only two tactics: lobbying and arousing public opinion. Neither of these had yielded results except in the heat of a strike. They would have to find a tactic which was as damaging to production as a strike but which would not impoverish their members or recreate rifts among the locals.

ENDNOTES CHAPTER IV

1. Charles Mannheim, *De la condition des ouvriers dans les manufactures de l'Etat (Tabac-Allumettes)*, Paris: 1902, 422- 424.
2. ibid., 298.
3. ibid., 435.
4. Socialist deputies who were favorable to the match workers in 1890 were: Eugene Baudin (Bourges), Antide Boyer (Bouches-du-Rhône), Jean-Baptiste Dumay (Seine), Antoine Ernest Ferroul (l'Aude), Emile Goussot (Seine), and Cristophe Thivrier (l'Allier). Among their non- socialist allies were Jules Faillard and Gustave Mesureur. Gaillard, deputy from the Oise, was a republican who worked in favor of social security, insurance, and mutual aid societies. Mesureur (Seine) was a radical-socialist who distinguished himself by working for the *prud'hommes* councils (industrial arbitration), for free employment agencies, for the reopening of the *Bourse du Travail*, and against the repression of union activities. In 1893 these socialist deputies, all from the Seine, were added to the match workers' allies: Marcel Sembat, Edouard Vaillant, Albert Walter, Emmanuel Chauvière, René Chauvin, Prudent Dervilliers, and Armand Rouanet.
5. Paul Pic, *Traité élémentaire de législation industrielle, Les lois ouvrières*, 3rd edition, Paris: Arthur Rousseau, 1909, 463.
6. Karl Marx, ibid., 480.
7. *Le Temps*, May 17 and 18, 1888 and France, Ministère du Commerce, de l'Industrie, des Postes, des Télégraphes, Office du Travail, *Les Associations professionnelles ouvrières*, Volume I, Paris: Imprimerie Nationale, 1899, 555.
8. At the organizing meeting of the Le Mans tobacco factory in 1892, the prefect presided; at the Tonneins factory in the same year, the mayor sat on the podium. Mannheim, ibid., 302.
9. A Madame Ménard was active in the Trélazé local. The first slate quarriers' union in Trélazé was organized in 1880 by Ludovic Ménard, who led a six-week strike in 1891 and was under police surveillance as the "uncontested leader of the anarchists at Trélazé." He went on to become the secretary-general of the National Federation of Slate Workers from 1904-1906 and the secretary-general of the Angers *Bourse du Travail* from 1918-1921. The slate workers had a history of democratic sympathies and militant action dating back to 1848 when they first struck over wages. In 1855 they marched on Angers in the hopes of

proclaiming the democratic and social Republic. Arrests and deportations had not changed their character. Maurice Poperen, *Syndicats et luttes ouvrières au pays d'Anjou*, Angers: 1964, 10-23; Jean Maitron, ibid., volume 14, 62; Archives départementales du Maine-et-Loire, 71M4.

10. Mannheim, ibid., 261 and 298-302.

11. By March 1893 the Federation had elected their own secretary-general Ernest Deroy, and a central committee of six men. Deroy was 30 years old and was living with or married to a 29-year old machine operator with a seven-year old son. Of the other central committee members, we know that Paul Bordat was 31 years old, had worked at a match factory since at least 1890 and was married to a match worker. Jacques Aschbacher was 30 years old, had started work in the Paris factories in 1880 and was also married to a match worker. These men were typical of the leadership of the union through the 1890s in two ways. They were in the prime of life and they represented families of match workers. Spouses, brothers, sisters, and parents were often co-workers, which was an added reason for solidarity. Archives de la Préfecture de Police de Paris (APP), 1.408 bis and personnel dossiers of Pantin-Aubervilliers examined at the factory at Saintines.

12. "From the moment when the Pantin-Aubervilliers workers founded the Match Workers' Federation, there was no end to agitation in the workshops and numerous acts of insubordination took place." Mannheim, ibid., 436.

13. Average daily wages in 1893: four francs seventy-seven centimes for men, three francs twenty-five centimes for women. *Bulletin de l'Office du Travail*, "L'Exploitation du Monopole des Allumettes Chimiques pour l'Année 1894", Paris: Imprimerie Nationale, 1896, 179.

14. Direction Générale des Manufactures de l'Etat, *Compte en Matières et en Deniers de l'Exploitation du Monopole des Allumettes Chimiques pour l'année 1894*, Paris: Imprimerie Nationale, 1895, vii.

15. Mannheim, ibid., 419-426.

16. *Le Temps*, March 20, 1893.

17. "L'Echo des allumettiers", Bègles, January 1898.

18. Mannheim, ibid., 424, 426; *La Marseillaise*, April 14, 1893.

19. Administration circulars from the years 1882-1901 do not contain such instructions on discipline until the October 7, 1900 circular

#62 on union-management communications and the October 30, 1900 circular #63 on the procedure for filing grievances and the number of workers who will be received as a delegation at each level of management. Archives of the factory at Saintines, Direction Générale des Manufactures d'Etat, *Circulaires 1882-1901*.

20. Workers' benefits in francs 1890-1892

Year	Francs
1890	52,012
1891	88,754
1892	102,490

The Administration was favoring a change in consumers' tastes by lowering the price of red phosphorous safety matches by 20%. However, there were unknown quantities of contraband white phosphorous matches in circulation. The spokesman who claimed that there had been only one case of necrosis since 1890 was wrong. The State had recorded five cases, two of which occurred at Pantin and one at Aubervilliers. However, the State could claim that workers were safer than they had been under the private company. The private company had reported an annual average of 5.75 cases or a total of 23 cases from 1886-1889. The State record for its first three years was 1.67 cases per year. Direction Générale des Manufactures de l'Etat, *Compte en Matières et en Deniers...1894*, ibid., iv, ix, 108, 109 and *Le Temps*, March 22, 1893.

21. The police investigated Deroy's background in order to have something to use against him. At the age of 16 he had spent six days in jail in Evreux for begging and 15 days in jail at Marseille for vagrancy. A few months later he had been sentenced to three months in a Montpellier jail for begging again. Upon his arrival in Paris at age 17, he had received a six-month sentence, which was reduced to two months, for breach of trust. However, once he settled in Paris he did nothing which could have given the police reason to disturb him. At the 1891 census, the 27-year old Deroy was a match worker and was living with a seven-year old son who bore the name Deroy and with the child's mother. It is unclear whether they were married. At the 1896 census he was living with the son, a daughter and their mother. APP 1.408 bis and AD Seine, D2M8 *Dénombrements* 1891 and 1896.

22. Mannheim, ibid., 437.

23. Antide Boyer had started work at the age of eight as a potter's
    apprentice. He was elected to the Marseille City Council in
    1884 as a socialist worker and to the Chamber of Deputies
    from 1885-1909. Among his first concerns in the Chamber
    was financial compensation for sick, injured, or aged workers.
    Jean-Baptiste Dumay had organized strikes at the Schneider
    Coal Mines in 1870, had known exile and amnesty, and had
    already served the match workers of Pantin-Aubervilliers as a
    City Councillor from Belleville from 1887-1890. Elected to the
    Chamber in 1889, he distinguished himself on workplace health
    and safety by proposing, in 1891, an international exposition of
    equipment designed to prevent industrial and mining accidents.
    Antoine Ernest Ferroul began his career as a doctor to the
    poor in Narbonne. An early socialist organizer, first of the
    Fédération des Travailleurs socialistes de France, then of the *Parti
    Ouvrier Francais*, he made himself known upon his election to the
    Chamber in 1889 as one of the deputies who could be counted on
    by workers. On May 1, 1889, a massive workers' demonstration
    was prevented by the cavalry from presenting their demands to
    the Chamber. Ferroul was one of the elected officials who was
    delegated in their place. He traveled throughout the country
    and spoke in the Chamber on behalf of workers from all parts
    of France: Fourmies in 1891, Carmaux in 1893. Therefore, it
    was not surprising that the match workers turned to him, despite
    the fact that there were no match factories in his *département*
    of the Aude. Emile Goussot represented a district including
    Pantin-Aubervilliers from 1890-1902. He was also one of the
    founders of the socialist group in the Chamber. He continued his
    interest in workplace health and safety after he lost his Chamber
    seat by specializing in his law practice in workers' compensation
    cases. Antoine Jourde, ex-umbrella factory worker and organizer
    of the *Parti ouvrier* since 1882, was elected to the *prud'homme*
    council of Bordeaux in 1885. He represented an area including
    the match factory town of Bègles from 1889-1902 and 1906-1910.
    Jean Maitron, ibid., volume 11. 34-3; volume 12, 102-103 and
    186-189; volume 13, 133-134 and Jean Jolly, *Dictionnaire des
    parlementaires francais*, Paris: PUF, 1960.

24. *Bulletin officiel de la Bourse du Travail de Paris*, Paris: Alle-
    mane, 1893, 594.

25. Federation representatives told *Le Temps* that two workers were

very sick at that time, but were not receiving compensation. Marguerite Beingen, aged 31, had been suffering from necrosis for 18 months, since October 1891. Jean Weber, aged 27, had lost both jaws and his upper palate to necrosis since August 1891. *Le Temps*, March 21, 1893.

26. *Le Rappel*, March 28, 1893.

27. *L'Intransigeant*, March 21-28, 1893; *Le Temps*, March 22, 1893; *Le Rappel*, March 23, 1893; *La Gironde*, March 22-28, 1893 and Office du Travail, *Statistique des Grèves et des Recours à la Conciliation*, Paris: Imprimerie Nationale, 1893, 351-358.

28. *Le Temps*, March 30, 1893 and *L'Intransigeant*, March 30, 1893.

29. AD Maine-et-Loire, 71 M3 and Mannheim, ibid., 440.

30. *Bulletin officiel de la Bourse du Travail de Paris*, ibid., March 26, 1893, 595; *Le Temps*, March 25-31, 1893 and Marseille, Conseil municipal, *Délibérations du Conseil municipal 1893*, Marseille: Imprimerie Moullot, 1896, 86.

31. *Le Socialiste*, March 26, 1893; *Le Temps*, March 24, 1893; *L'Intransigeant*, March 29, 1893.

32. *Le Temps*, March 29, 1893; *Le Rappel*, March 28, 1893; *La Justice*, March 23, 1893; *Le Socialiste*, March 26, 1893.

33. *Le Temps*, May 7, 12, 1893.

34. *Deuxième Congrès national des ouvriers et ouvrières des manufactures d'allumettes*, St. Denis: 1894.

35. At the time of the Second Congress, these were the most recent cases. Marguerite Beingen had suffered from necrosis from September 1891 to March 1893, had undergone three operations, and had returned to the Aubervilliers factory in 1894. Jean Billot had suffered since 1893; his fourth operation included the removal of his right cheek and the implantation of an artificial jaw. Marie Herff had been diagnosed as a necrosis patient in August 1893. She had had all of her teeth pulled, but a hole in her upper jaw indicated the advisability of a delicate operation under her eye. Lisa Magard had developed a split upper lip and damage to her upper jaw since 1893. Charles Pouteaux died in 1894 after his third operation due to his reaction to the anaesthesia. His widow received no compensation. Marie Schmitt had contracted necrosis in November 1893. Jean Weber, who had had his teeth and upper palate extracted, was declared cured in March 1893 and returned to work in a safe workshop. However, he needed an artificial jaw which cost 1,000 francs. *Deuxième Congrès national des ouvriers*

et ouvrières des manufactures d'allumettes, ibid.; *Le Temps*, April 5, 1895; *La Petite République*, February 24, 1895.

36. *L'Eclair*, May 27, 1894.

37. Some doctors believed that turpentine fumes were an antidote to white phosphorous. Ironically, turpentine is also a poison. Prolonged exposure causes headaches and irritability.

38. *Deuxième Congrès national des ouvriers et ouvrières des manufactures d'allumettes*, ibid.

39. Percentage of white phosphorous matches to total matches:

1892	86.4%
1894	81.0%
1895	77.0%

    Letter from Director-General Charles Favalelli to Minister of Finances November 25, 1895 in Direction Générale des Manufactures d'Etat, *Compte en matières... 1894*, ibid., iv, vii.

40. Journal Officiel, Chambre, *Débats*, March 14, 1895, 939. Amendment proposed to chapter 95: Wages and salaries of the Administration of State Manufactures by Marcel Sembat, Edouard Vaillant, Jean Jaurès, René Viviani, Eugène Baudin, Maximilien Carnaud, René Chauvin, Arthur Groussier, Alexandre Zavaès, Victor Dejeante, Gustave Rouanet, Emmanuel Chauvière, Pascal Faberot, Jules Coutant, and Léon Gerault-Richard.

41. ibid., March 15, 1895, 944.

42. Among those who made public statements of support were: City Council members Alfred Brard (19th *arrondissement*, Paris); and Joseph Fournière (18th *arrondissement*, Paris); Deputies Antide Boyer (Bouches-du-Rhône); René Chauvin (St. Denis); Emile Goussot (Pantin-Aubervilliers); Clovis Hughes (19th *arrondissement*, Paris); Ernest Roche (17th *arrondissement*, Paris); Albert Walter (St. Denis); Senator Maximilien Carnaud (Bouches- du-Rhône).

43. *L'Intransigeant*, April 20, 1895.

44. The contribution from *La Solidarité des Femmes* was presented by Citizens Potonié-Pierre and Vincent, feminist journalists.

45. *L'Intransigeant*, April 9, 1895.

46. ibid., April 9, 1895.

47. The central committee urged their fellow workers in the 41 major Belgian match factories to strike in solidarity. There is no evidence of an answer. *L'Intransigeant*, April 3, 1895.

48. *Le Rappel*, April 7, 1895.

49. Letter from the central committee to the membership as reported in *L'Intransigeant*, May 5, 1895.

50. idem

51. AD Maine-et-Loire 71 M4.

52. idem. Neither Louis (Jean-Pierre) Bridier nor Madame Ménard is mentioned in Jean Maitron, ibid. He and a woman named Bridier were among the delegates to the Director of Trélazé in September 1890 according to the Director's report to the Minister of Finances. Madame Ménard was mentioned for the first time as one of the recipients of this letter.

53. Paul Rondet, born in 1867, was listed as the president of the Pantin-Aubervilliers unions in a police report of April, 1893. He had worked as a match worker since either 1885 or 1889 and was married to a match worker. He is not listed in Jean Maitron,ibid. APP 1.408 bis.

54. AD Maine-et-Loire, 71 M4.

55. Neither man is mentioned in Jean Maitron, ibid. Halopé had been one of the delegates to the Director of Trélazé during a September 1890 strike and to the 1892 Congress. The police listed him as an anarchist in answer to a ministerial circular of September 16, 1893 and again in March 1895. Aubert's name appears for the first time in the police report on his and Halopé's trip to Saintines. AD Maine-et-Loire, 71 M2 and M3.

56. Maurice Poperen, ibid., 51-69. The protest was published in the syndicalist newspaper, the *Eclaireur de Tours*.

57. AD Maine-et-Loire, 71 M4. Report of the special officer, April 4, 1895.

58. AD Maine-et-Loire, 71 M4. Report of Officer Bouhier, commander of the Trélazé brigade.

59. AD Maine-et-Loire, 71 M4. Report of special officer.

60. idem

61. The president, Madame Mahoux, is mentioned for the first time as the signatory of this strike notice. She is not listed as a delegate to either the 1892 or 1894 Congress.

62. AD Maine-et-Loire, 71 M4. Report of special officer to prefect, April 2, 1895.

63. ibid., April 4, 1895.

64. idem

65. idem

66. Mannheim, ibid., 435.

67. In the 1896 census he is listed as a deliveryman.
68. AD Maine-et-Loire, 71 M4, April 4, 1895, Report of the special officer.
69. Villarmé or Villarmet and Henriette Bernard were mentioned for the first time when they were chosen as delegates. ibid., April 19, 1895.
70. ibid., April 27, 1895.
71. AD Maine-et-Loire, 71 M4.
72. idem and succeeding quotes
73. It is more likely that their deputies were simply uninterested. Jean Guignard was a republican who did not distinguish himself by strike support. Alexandre Fairé was anti-republican and generally indifferent to workers' causes.
74. *Le Temps*, May 2, 1895.
75. AD Maine-et-Loire, 71 M4, Letters marked confidential from Director Sévène to Director-General, May 2 and 3, 1895.
76. AD Maine-et-Loire, 71 M4.
77. idem
78. *L'Intransigeant*, May 6 and 9, 1895.
79. AD Maine-et-Loire, 71 M4
80. Direction Générale des Manufactures de l'Etat, *Compte en matières...1894*, ibid., xi.

CHAPTER V: THE UNION STRUGGLE 1896-1898: CONTINUING THE FIGHT BY OTHER MEANS

From the summer of 1895 to 1898 there ensued a series of Federation press announcements that a new match had been discovered: it was absolutely non-toxic, inflammable against all surfaces, contained no white phosphorous, and cost the same or less than the poisonous match.

However, the scientific committee, which had been called by the Minister of Finances Ribot in April 1895, was more cautious about each of the alternative formulae.[1] In July they reported that there was a problem with the existing red phosphorous strike-anywhere match: a tendency to ignite with a little pop. The Ministry of Finances also claimed that the existing equipment was not suitable to produce that type of match. There was another safe formula which was rejected at the same time because it produced matches which were difficult to light. This announcement was greeted with mockery by the republican socialist editor of *l'Intransigeant*, who claimed that the new matches could hardly be worse than the Administration's usual product which "six times out of eight" did not light.[2]

This criticism was an oblique reference to one of the underlying problems in the whole match controversy. Everyone knew that the State monopoly produced poor quality matches and that competition from contraband was still fierce. Therefore, more seriously than in 1889, Parliament and the Ministry of Finances now considered an attack on contraband in order to make a ban on white phosphorous financially possible.

The efforts to survey the commerce in phosphorous before 1872 and the commerce in illegal matches sinces 1872 had been unsuccessful.[3] As we have seen, the *Compagnie Générale* had not succeeded in pushing illegal production out of the market with its low quality matches.

Ironically, even after direct administration began in 1890, the government did not enforce its own anti-contraband regulations of 1884 by building circular walks around all of its factories and covering all windows with bars. In fact, under direct administration there is evidence of less success in stopping contraband every year until strong decrees were issued during and immediately after the strike of 1895.[4] In that year a ten-franc bounty for the capture of illegal dealers was added to the 1875 law. There were also fines of 100-1000 francs plus imprisonment for the manufacture or transportation of matches. Even possession of the tools or empty match boxes was cause for arrest. The parents of minors who broke the law were to be tried as codefendants. As one indication of the trifling but persistent nature of contraband, the 1895 decrees provided that individuals who were arrested for less than fifty francs of illegal matches could be tried in group trials.

Results were not obtained until 1897 (See Chart XX: Bounties for Seizure of Illegal Matches and Increase/Decrease over Previous Year 1892-1897). This indicates that although the new anti-contraband measures went into effect in the summer of 1895, they did not prove satisfactory for eighteen months. During those eighteen months something else happened to increase the urgency with which the government viewed control over the commerce in matches.

CHART XX: BOUNTIES FOR SEIZURE OF ILLEGAL MATCHES AND INCREASE/DECREASE OVER PREVIOUS YEAR 1892-1897

Year	Bounties (francs)	Increase	Decrease
1892	67,191		
1893	60,890		6,301
1894	55,749		5,141
1895	51,455		4,294
1896	45,864		5,591
1897	54,247	8,383	

Data from Direction Générale des Manufactures de l'Etat, *Compte en Matières...1898*, 1898, 65.

That new urgency grew out of the Federation's continuing efforts to enforce every rule in their favor and to expand their rights as state workers without going so far as to declare another strike. They did this by going over the heads of their factory directors to lobby their deputies and the Minister of Finances. At the same time, they ran a successful press campaign, thus keeping the public aware of the responsiveness or non-responsiveness of these republican officials. In all these actions, the workers at Pantin and Aubervilliers acted as if they were all of the match workers of France. The press treated them accordingly. The members of the other locals were either silent or happily agreed with the actions of their Parisian comrades.

Barely two months after the strike, the Pantin-Aubervilliers workers protested a work rule first to their Director, Brandeis; then to the Director-General of State Manufactures, Favalelli; then to the Minister of Finances, Maurice Ribot. By July 13, 1895, they were demanding that the victims of phosphorism and necrosis from Marseille and Trélazé be brought to Paris for medical care from Dr. Emile Magitot, that the sick from all seven factories be sent to the country to recuperate and that all necrosis sufferers receive free milk, which was believed to be an antidote. Director-General Favalelli answered that country air and milk would be provided, if necessary, on a case-by-case basis, but that provincial doctors were sufficiently qualified.[5]

Next, a delegation from the Federation visited Minister Ribot on September 18, 1895. He showed enough concern to make a personal visit to Pantin- Aubervilliers. There he told the match workers that samples free of white phosphorous had arrived. He agreed with them that their buildings were urgently in need of repairs for which he would ask a million francs. A liter of milk per day would be provided to each necrosis sufferer who was on sick leave. He shook hands, spoke respectfully to the workers, rescinded some punishments and impressed them favorably with his concern.[6]

The Pantin-Aubervilliers leadership occupied public attention through the press for the next three weeks over the promised milk. The readers of *l'Intransigeant* were treated to an almost daily serial on the match workers' milk controversy from October 5 to 24. *La Libre Parole*, *La Lanterne*, *La Petite République*, *Le Radical*, and *La Justice* with a combined circulation of more than 293,000 also covered the story.[7] The spokesmen for the Pantin-Aubervilliers locals claimed that Minister Ribot had spoken of three liters per person per day but had provided only one. They also complained because the milk was distributed at the factory, half a liter in the morning, another half-liter in the evening. Workers were required to drink it on site. This was inconvenient if not debilitating for those who had to leave a sick bed. Also, the requirement of drinking it on the premises suggested that the Administration did not trust the workers to use the milk for their own consumption. Furthermore, the Secretary-General of the Federation accused the Administration of buying only 40 liters per day although there were 80 necrosis victims. He insisted on a milkman for home delivery. On October 5, 1895 the Pantin-Aubervilliers local announced that its members would refuse the milk until it was delivered to their homes.

Les Débats, a center-left publication, took notice of the contro-

106

versy to express annoyance:

> "One begins to find, not without reason, that these State workers occupy the public a bit much with their actions."[8]

One conservative paper carried a satirical column, on October 14, 1895, suggesting that the workers wanted not only home delivery of the milk, but also sugar, rum, and warm croissants. Furthermore, the article cited a hypothetical single male match workers who wanted a pretty milkmaid.[9] But on the following day, the Pantin workers voted their thanks to the press and to the Minister for their bowl of milk delivered to their homes.[10]

Again the Pantin-Aubervilliers workers had won without calling a strike. However, the Administration, recognizing the financial drain that necrosis represented, started to look into automation as well as non-toxic matches. By March 1896 there was discussion of whether the Administration would not rather buy land at Le Bourget outside of Paris. There they would build a factory with the enclosed machinery available from the Diamond Match Company of the United States.[11] This new factory would replace five of the existing factories and eliminate two-thirds of the workers.[12] The mayors and town councillors of the cantons around the town of Saintines (Oise) took the threat seriously enough to send a plea in April 1896 to the Minister of Finances. They argued for keeping their factory open for three reasons. First, their area was healthy:

> "for six years there have been no cases of necrosis at Saintines, which has an exceptionally healthy climate..."

Second they had access to wood in the Compiègne forest. Third they were willing and able to accept the American machines:

> "the buildings which are currently used for wood-cutting are spacious enough to accomodate a certain number of American machines,"[13]

A delegation from the Federation conferred with the socialist deputies who represented the districts where Pantin-Aubervilliers workers lived: Emmanuel Chauvière, Prudent Dervilliers, and Albert Walter. A mass meeting of 550-800 people was held. Journalists from *Les Débats*, *l'Agence nationale*, and *l'Agence Havas* press services attended as well as the usual socialist and radical press. The best that Deputy Walter could promise was an amendment to the most recent finance bill. Walter expressed the hope that in this way he could secure other State jobs for displaced match workers. Minister

Ribot, who had held out so much hope to the match workers at the time of his personal visit to Pantin-Aubervilliers, had lost his office to Paul Doumer. Doumer, a left radical, at first impressed the match workers as a friend of labor. Like his predecessor, he made a visit to Pantin-Aubervilliers where he was overheard to murmur that indeed, the factories were too filthy to even be used as prisons. He too promised the speedy substitution of an unspecified safe formula, and even rehired a match worker who had been thrown out for stealing two boxes of matches. But this *confrère rose*, as the editor of *l'Intransigeant* called him, was also soon considering automation.[14]

In May 1896 Georges Cochery became the Minister of Finances. He did not even answer the match workers' requests for meetings. According to the radical daily *La Lanterne*, nothing good could be expected from the Méline cabinet.[15] The match workers took to publishing their letters to Minister Cochery in the left press in the disgruntled hope of at least keeping public opinion aroused.

Inadvertently, company doctors provided the next tactic of class struggle: the sick-out. As we know, there was already dissatisfaction among the workers over excessive tooth-pulling. At the April 1896 congress, the delegate from the newly-organized local at Aix complained:

"...the dentist shouldn't use poor men and women workers for experience because he's new at his trade. It's very childish (*puéril*) that a tooth which hardly has a cavity and has never hurt its owner should be pulled. It could well happen, whether from ill will or from any other cause, that the worker would develop a necrosis..."[16]

Both workers and company doctors were increasingly unwilling to wait for control over the disease, but workers were increasingly unwilling to accept what amounted to preventive mutilation. When doctors could not oblige or convince match workers to have their teeth pulled, doctors could still send workers away to rest.

For this reason, the list of workers on sickleave from Pantin-Aubervilliers rose starting in 1894 and resulting in a crisis in 1896. In December 1894, Pantin-Aubervilliers had 32 necrosis sufferers on paid sick leave. By December 1895, the number had risen to 123. By August 1896 it was 226 or 36% of the work force. At this point, the Minister of Finances appointed the medical commission which had been demanded by the Academy of Medicine two years earlier.[17] The doctors' tasks were to respond to the immediate crisis at Pantin-Aubervilliers and to offer long-term suggestions. The Minister of Finances wanted to know whether match manufacturing was becoming more dangerous than ever or workers were collecting sick pay illegitimately. If match manufacturing were becoming more dangerous despite the hygiene measures, perhaps there was something wrong with the hygiene measures. Another alternative was that the workers or the company doctor were misusing the system.

In December 1896, the medical committee reported. Of the 238 workers on sick leave (the number had risen between August and December) 142 were judged to be unfit to return to work. Of those, 125 had such bad teeth or such generally weak constitutions that the doctors considered them to be high risks. Another five workers were put on full pensions for other reasons. Only twelve workers were diagnosed as suffering from phosphorism or necrosis (1.7% of the Pantin-Aubervilliers workforce). All 142 would receive compensation according to their age and the length of their service. The others were declared to be malingerers and were invited to return to work immediately.[18] As the remaining workers were still to handle white phosphorous and were not reassured by the vague compensation offer, they continued to protest. In March 1897, at a meeting of 1200 people, the Pantin-Aubervilliers locals voted unanimously to refuse any further dental exams until they were sure that the exams would not lead to a financially uncertain involuntary retirement. This refusal became the issue of the summer of 1897.

On August 5, 1897, 30 workers at Pantin were selected for dental exams. Only ten submitted. The refusees were fired. The next day, the fired workers were barred from the building by no less than three people at each door: one supervisor and two policemen (one *gendarme* and one *gardien de la paix*). When suspicious fires broke out in the drying rooms, a supervisor was placed in each one. The socialist daily *La Petite République* accused the Administration of a military occupation of the factory.[19] The article also claimed that the dental exams had been suspended for the previous ten months and went on to suggest that the only reason for reinstituting them at that time was to replace permanent workers with two-month temporaries. Because of the possibility of another strike, an unusual number of newspapers covered the next union meeting on August 7, 1897: *l'Evènement, La Liberté, l'Intransigeant, La Libre Parole, Le Dix-Neuvième Siècle, Le Prolétaire, Le Temps,* and *l'Echo de Paris* (combined circulation more than 182,000).[20] There it was decided not to strike. Instead, the workers would submit to the dental exams only on the condition that anyone who was found to have a necrosis would be rehired when cured. In fact, the union had backed down and returned to work without a guarantee. However, the refusees were allowed back and peace was momentarily reestablished.[21]

By this time the American machines were no longer under consideration because the Diamond Match Company had placed too high a price on its patent.[22] This lack of capitalist solidarity permitted the match workers, who still did not have the strength or unity for another long strike, to continue to exert pressure. While the Administration engineers tried to duplicate the American machine (they succeeded in 1900), the Federation's representatives testified before the Parliamentary budget committee that even if enclosed machinery were introduced, that would not still their members' protests. They explained that as long as white phosphorous was in use, certain functions would still expose them to the poison: the preparation of the chemical mix, the cleaning and repairing of machines, and the moving of the finished product to the stockrooms.

Indeed, as the directors of the French match industry were not in the position of the American or British manufacturers, who could simply fire necrosis victims or hide their cases,[23] the French had to find a solution which was satisfactory to the union. Unlike the British match workers, whose union was weak, and who were in the private sector of the economy, the French match workers were in a position to bleed their employer financially and embarass their employer politically until the poison was banned. The only alternatives for the Ministry of Finances were either to produce nothing but Swedish safety matches or to find a substitute. The government was not willing to convert entirely to safety matches because the fight against contraband white phosphorous matches had still not shown results by the summer of 1897. Also, the equipment for Swedish safety matches was more expensive than the equipment for white phosphorous matches.[24] Therefore, the Ministry of Finances was keeping in mind the wastefulness of spending on new equipment if the invention of the alternative match was about to occur.

Reports on the development of alternative matches punctuated this period of 1896-1898, during the ministerial changes, the milk controversy, and the sick leave investigations. For example, in January 1897, a union delegation offered non-toxic samples to the *commission du travail* and saw to it that *La Lanterne* and *l'Intransigeant* reported it. The matches were probably from the inventor Pouteaux who seemed to be using the Federation and the press to lobby the government in favor of his invention. On several occasions, police informers reported on union votes in favor of "Pouteaux", "Touteau" or "Ruteau" matches.[25] Why then did the Administration reject them? *Le Figaro* provided the answer:

> "The only defect of the permanganate of potassium match is that it leaves a blackish smudge on the surface it rubs against. This is a bit bothersome for pearl-grey trousers. But, is it necessary to always put the seat of one's pants to this familiar purpose?"[26]

Apparently, this familiar purpose **was** sufficient reason. After all, those who bought white phosphorous matches were not only the wearers of pearl-grey trousers but also working people, artisans, and shopkeepers who wanted to light a match with only one hand against whatever surface was convenient. Therefore, the Administration decided that marketing a dirty match instead of an unreliable match was not going to increase their share of the market.

In December 1896, *Le Petit Parisien* announced that a second alternative match had been found by Emile Bert, an engineer at the *Ecole des Arts et Manufactures*. It could be produced with existing equipment. Bert had discovered the formula in October 1895, run tests at Aubervilliers from December 1895 to February 1896 and left 3000 boxes there for the Ministry of Finance's approval. In October 1896, still having no answer, he inquired to the budget committee about the delay. The Federation saw to it that the press was informed of the government's sluggishness.

A third match came from an inventor named Otto Miram who wanted 300,000 francs for his patent. A Federation delegation took the initiative of traveling to Westphalia for samples which they brought to the Minister of Finances.[27]

In January 1897, the Minister assembled his committee of scientists to examine Pouteaux's, Bert's, and Miram's matches. The Minister authorized trial production of Pouteaux matches in September and of Miram matches in November. First, the Federation claimed bad faith on the part of the management at Pantin during the tests. For the duration of the tests, management had changed the pay system from piece rate to a daily rate. The union leadership interpreted this change, which they had so desired earlier, as an effort to decrease productivity and then blame the decrease on the Pouteaux matches. Then,

> "These gentlemen (management) walked through the work-
> shops, took the matches, broke them instead of lighting them
> and declared that they were worthless. Behind them the
> workers picked up the same matches and lit them with no
> trouble at all."[28]

Next, the Federation suggested a market trial of these two
matches *and* a simultaneous withdrawal of all other matches in a
given area to see how the consumers reacted. Instead of following
the Federation's suggestion, in December 1897 the Administration
marketed them in competition with white phosphorous matches,
without advertising, and then claimed that the public had not
appreciated them. The Federation also protested that Otto Miram's
"Triomphe" matches were priced the same as white phosphorous
matches although their production cost was 30% less. Several
deputies who were sympathetic to the match workers complained that
either the new matches had not been available in their *département*
or no one had known to look for them.[29]

After two more trials by Pouteaux in March and July of 1898,
one more by Miram in June of 1898, and one by the State engineers
Sévène and Cahen at the Trélazé factory in March of 1898, the
Administration settled in October on phosphorous sesquisulphide
matches.[30] The successful inventors were the State's own industrial
engineers Sévène and Cahen.

In the next few months the press indicated no particular consumer
reaction to the new match rather than the rejection which the
Administration had so feared. In the next few years the profit figures
showed no significant change (See Chart XIX: Profits 1890-1902 in
the previous chapter). The Administration's extreme caution about
imposing the ban had been unrealistic. While the Administration had
stalled, workers had been mutilated and the Federation had learned
how to exert its strength and maintain its unity without calling a
strike.

ENDNOTES CHAPTER V

1. On April 25 and 30, 1895, the scientific committee examined alternative matches. The members of the committee were: Senator Berthelot, member of the Institute, Perpetual Secretary of the Academy of Sciences; Schloesing, member of the Institute, Director of the Applied School of State Manufactures; Troost, member of the Institute; Soirau, member of the Institute, Director of the Central Explosives Laboratory; Vieille, Explosives Engineer; Sollier, Engineer of the State Manufactures, Bureau chief to the Minister of Finances. *Le Temps*, April 6, 1895.

2. *L'Intransigeant*, July 6, 1895.

3. A statute of October 23, 1846 regulated the sale, purchase, and use of dangerous substances. A decree of July 8, 1850 specifically listed phosphorous among those substances. After the State monopoly went into effect, a circular of January 20, 1875, from the Ministry of Commerce to all police prefects reminded them to survey the commerce in white phosphorous. By that date no match manufactures other than those controlled by the *Compagnie Générale* had any reason to buy white phosphorous. However, manufacturers of red phosphorous and some explosives plants still had a legitimate need for the product.

4. On March 22, 1895 the Chamber of Deputies passed a budget including higher fines. The decrees referred to were found in Georges Paulet, *Revue du Commerce et de l'Industrie*, April 16, 1896, 208-209. More precise regulations were in the decree of July 19, 1895, idem, 366-369. Ironically, this decree posed obstacles to inventors who wanted to win the prize for the safe match.

   Article 1: Within ten days of the promulgation of this decree, all holders of phosphorous will make a declaration of quantity to the Administration.

   article 2: Within the same time period manufacturers of phosphorous must declare themselves to the Administration with a description of the factory indicating the daily and hourly schedules. An identical declaration must be made for all new factories one month before beginning production.

   Article 3: Any change in daily or hourly work schedules must be declared 48 hours in advance. This includes stopping, suspending, or starting up work.

   Article 4: The Administration may require that two chairs and a table with a locked drawer be at the disposition of its employes in a suitable place within the factory. (Clause on rental of this furniture).

Article 5: The manufacturer may not mold phosphorous in workshops other than those indicated in the declaration specified in article two.

Article 6: It is forbidden to stock finished phosphorous – ordinary or amorphous – in places other than those indicated in the aforementioned declaration.

Article 7: The tanks with condensing apparatus in each oven must be numbered, so should the purifying installations.

Article 8: As soon as work with ordinary and amorphous phosphorous is completed, the phosphorous will be placed into closed boxes, then immediately into numbered containers. Wastes from the molding process must be placed, gradually throughout the workday, in one place in the workshop. At the end of the workday, all these wastes will be weighed and entered into the manufacturer's register, then carried to the remodelling area.

Article 9: The manufacturer of phosphorous must keep a separate record without gap, without crossing out or adding on in the course of operations:

1. the quantities of raw phosphorous removed from the condensers and carried to the purification workshops
2. the quantities of purified phosphorous carried either to the molding workshop, the amorphous phosphorous workshop, or the workshops for phosphorous-derived products. The notation will include the number of the batch, the number of the purifier, the number of receptacles for transport, the waste from these receptacles and the weight of raw or purified phosphorous they contain.

Article 10: The Administration will provide a second register to the manufacturer, who will record at the end of every day:

1. the quantities of ordinary phosphorous molded, packed, and carried to the storeroom
2. the quantities of amorphous phosphorous packed and carried to the storeroom. In the molding and amorphous phosphorous workshops, an employe of the manufacturer will enter in a register the quantities of phosphorous packed into containers. The entries in this register will be repeated at the end of the workday by the manufacturer who will enter in his register the weights and numbers of each of these containers as well

as the total quantity.

Article 11: Any phosphorous found outside the workshops or the storerooms designated for its manufacture, storage or processing will be seized.

Article 12: The quantities of manufactured phosphorous, ordinary and amorphous, in stock at the time of the entry into effect of this decree and those quantities introduced afterwards will be verified by the employes responsible for an account of the storerooms. By the same token, quantities sent out for export or domestic consumption and brought back to the factory will be noted. Quantities which leave the factory as regular outlays will be marked off. The administration can mark off quantities of altered phosphorous which must be brought back for reprocessing.

Article 13: The employes may at any time examine the accounts of the storeroom of manufactured products. If the account shows an excess or a loss, the excess will be seized and added to the accounts. In either case, a written account will be made for the application of the penalties according to article five of the law of September 4, 1871.

Article 14: Any person wishing to buy or sell phosphorous must make a preliminary declaration to the Administration above and beyond the formalities required by the ordinances of October 29 and November 6, 1846. He will receive an "ampilation" which will serve as authorization. Vendors of phosphorous may not receive shipments except as regularly scheduled and may not sell on the domestic market except to authorized merchants or buyers who have satisfied the conditions of the article above.

Article 15: Anyone, whether manufacturer, chemist, or other, who wants to use phosphorous, must fill out a declaration at the mayor's office of the quantities desired and the intended use. A copy of this declaration, certified by the mayor, must be presented in duplicate to the director or the assistant director of taxes who will affix his stamp. A list will be kept by the buyer to be presented to employes of the tax bureau in charge of the use of phosphorous. Another copy will be sent by the buyer to the seller who will add it to his sales receipt.

Article 16: No phosphorous may be in circulation for domestic consumption or export except in lead-lined numbered containers or boxes available from the Administration or the Customs Bureau and accompanied by a permit. This permit will list the numbers and weights of each of the containers in the shipment. The distribution of permits for domestic use is subject to the presentation of a certified copy of the declaration mentioned in article 15 above. In case of discrepancies from the certificate of loading, or in the case of excess or loss noticed upon arrival, a report will be prepared for the penalties indicated in article five of the law of September 4, 1871 within the regulatory delays.

Article 17: Any phosphorous merchant or buyer of any quantity which will not be used immediately will open an account under the same conditions as manufacturers. Merchants and buyers who do not receive quantities greater than 100 grams of phosphorous within one year are not required to keep an account. This account will register quantities which exist at the time this decree goes into effect and those quantities which are received afterwards with permits. Among the outlays will be noted the quantities which are regularly sent out and the quantities whose use on the premises will be justified.

Article 18: An opinion from the *comité consultatif des arts et manufactures* will determine the alteration procedures to be used in agricultural or industrial uses of phosphorous before exit from the factory and after purchase.

Article 19: Articles 235, 236, 237, 238, and 245 of the law of April 28, 1816 and article 24 of the law of June 21, 1873 will be in force for verification by employes of the tax bureau in factories and at all other places where phosphorous is held.

Despite this decree, as late as 1909, contraband was still a problem. The following article was included in the law on finances: "Breaches of the laws and regulations on phosphorous are punished by the payment of a sum double the value of the matches which might have been produced, calculated at the rate of 1,000 francs per kilo of phosphorous manufactured, held, sold, or illicitly transported in addition to the penalties presently mandated." J.O. *Lois et décrets*, December 1908, 8971.

5. *Le Rappel*, July 20, 1895. The Marseille workers had probably initiated the demand that all match workers receive the attentions of Dr. Magitot. In September 1895 a Marseille match worker named Rosine Cayol received a subsidy of 150 francs to travel to Paris for the attentions of Dr. Magitot. Marseille, Conseil Municipal, *Délibérations*, Marseille: Moullot, 1895, meeting of September 14, 1895, 224.

6. *L'Eclair*, September 28, 1895.

7. Circulation figures from AN F18 2365 cited in Claude Bellanger, et al., *Histoire générale de la presse francaise 1871-1940*, Volume III, Paris: Presses universitaires de France, 1972.

8. *Les Débats*, cited in *L'Intransigeant*, October 7, 1895.

9. *Le Siècle*, October 14, 1895.

10. *La Petite République*, October 15, 1895.

11. The Diamond Match Company of the United States had an entirely enclosed conveyor belt process which, they claimed, had ended the problem in that country. Diamond Match's factories in Liverpool were also using that machinery.

12. *Le Matin*, March 17, 1896.

13. AD l'Oise, #105 Saintines, "Procès-Verbal de la Réunion du 12 avril, 1896."

14. *L'Intransigeant*, November 23, 1895.

15. *La Lanterne*, June 3, 1896.

16. *Troisième Congrès National des Ouvriers et Ouvrières des Manufactures d'Allumettes*, Marseille: April 11-21, 1896, 69.

17. Members of the medical commission were Drs. Emile Magitot, Fréderic Monod, and Emile Vallin. Magitot's biography has been discussed. This was his opportunity of a lifetime. According to the *Enciclopedia Universal Ilustrada*, phosphorous necrosis or *osteo periostitis alveolodentaria* also known as *periodontitis expulsiva* is "Magitot's disease." Dr. Monod wrote his thesis on bone disease, wrote for the Société anatomique and Société de chirurgie, was Hospital Surgeon at the Hospitals of Lourcine, Cochin, and the Maison royale de santé, collaborated on the *Dictionnaire des études médicales pratiques*, the *Nouvelle bibliographie médicale*, the *Revue médicale*, and *Transactions médicales*. In 1891 he was named Honorary Surgeon of the Hospitals. Dr. Emile Vallin, a military doctor, was a member of the Academy of Medicine and editor of the *Revue d'hygiène publique et de police sanitaire*.

18. Dr. Emile Vallin, "L'intoxication phosphorée et les allumettiers de Paris", *Revue d'hygiène publique et de police sanitaire*, January 1897, 94.

19. *La Petite République*, August 8, 1897.

20. Claude Bellanger et al., ibid.

21. APP 1.408 bis, August 9, 1897.

22. *La Petite République*, October 28, 1896.

23. In Britain, Diamond Match automated its factory at Liverpool, but the Bryant and May Company ran a dirty factory in London. In 1892 *The Star* revealed that one of the workers who was suffering from necrosis had been kept at home on paid sick leave with paid doctor's visits but denied her job when she was cured for fear that her mutilated face would arouse the other workers' anxiety. This incident and others led to a Home Office ruling that necrosis cases be reported to H.M. Inspectors of Factories. Another exposé by *The Star* in 1898 revealed that necrosis cases were still being hidden. There was a public hearing and a fine. This incident points up the difference between the British and French cases. The following explanation by a historian of the British match industry would have been equally true for the French if not for the strength of the French Match Workers' Federation: "In effect the (British) government was unwilling to cause economic disruptions in a small but important industry where the probability of suffering from a work-related disease seemed to be relatively low." Lowell J. Satre, "After the Match Girls' Strike: Bryant and May in the 1890s," *Victorian Studies*, Autumn 1982, volume 26, number 1, 7-31.

24. "At present, the installation cost for the most modest competitive machinery for the production of Swedish matches is 80,000 marks in round figures. For that reason, small scale factory or cottage production has been forced out of this branch of the match industry." Association internationale pour la protection légale des travailleurs, *Les industries insalubres*, Iena: Fischer, 1903, xxiv.

25. APP 1.408 bis. Report by "Félix", March 30, 1896.

26. *Le Figaro*, March 17, 1896.

27. "L'echo des allumettiers", January and June, 1898; APP 1.408 bis, January 26, 1897.

28. "L'echo des allumettiers", February 1898.

29. idem

30. This compound was richer in oxygen than white phosphorous; therefore, it was not able to attach itself to the oxygen breathed by the workers. For this reason, it would not fasten onto teeth, the necessary first step in the development of necrosis.

CHAPTER VI: THE LIMITS OF THE MEDICAL PROFESSION

"This industry has not ceased to be a preoccupation of the Council (of Public Health), in particular from the point of view of the workers' health."[1]

This opinion, from the Public Health Council of the Seine, was typical of the worried but impotent opinions issued by that council and its provincial analogues for the 50 years from the inception of the match industry until the government takeover. Why were the public health councils consistently ignored? And why were the illustrious doctors in the Academy of Medicine, the height of the profession, also ignored despite their repeated calls for the substitution of safety matches?

To understand French society's more than half century of indifference, we must realize that the medical profession as a whole had much less prestige and influence than we have come to expect in the twentieth century. Medical practitioners were not even unified by common education until the law of November 30, 1892, which outlawed *officiers de santé* or health officers. These men had fewer years of training than doctors and were authorized to work in only one *département*.[2] Doctors also looked down on dentists as mere mechanics. There was no professional organization which could speak for the majority until *l'Union des syndicats médicaux francais* formed in 1922. In addition, the period of time when white phosphorous was under investigation, from the 1840s to 1898, was a period when neither society nor doctors were accustomed to the medicalization of the industrial workplace. The prevailing liberal ethos ran contrary to any interference with the capitalist's right to hire whom he pleased and run his factory as he pleased. Even if there had been more than a few crusading, socially active doctors like Louis-René Villermé, Théophile Roussel, and Ambroise Tardieu, it is unlikely that their cries for government-imposed industrial safety and the protection of working class women and children would have been heeded.[3]

Perhaps most significant is that doctors had not yet acquired what the sociologist Haroun Jamous described as that:

"social and moral importance of technico-charismatic power founded on competence."[4]

Indeed, late nineteenth century doctors were not competent to correct many ailments and French doctors lagged behind even their British and German colleagues. Most doctors did not command social respect because they came from the lower ranks of the bourgeoisie and the gentry; thus, they were not rich and their family names were not known. Many doctors were the overworked and underpaid contractees of mutual aid societies whose members held open- ended contracts for any and all treatment at a fixed premium. Until 1872 any doctor outside of Paris had no ongoing contact with a medical faculty; there were none. Had he wanted to keep abreast of new developments, he could have subscribed to publications such as the *Annales d'hygiène publique et de médecine légale* (later the *Revue d'hygiène publique et de politique sanitaire)* which began publication in 1829, but even its authors were unable to suggest more than cleanliness, isolation, and ventilation as antidotes to most public health problems. Therapeutic nihilism and a respect for theorists rather than practitioners persisted through the nineteenth century. Too often the authors of medical articles contented themselves with reciting the history of previous opinions on a problem; rarely was experimental or large-scale statistical information provided. For example, in 1856 Dr. Ambroise Tardieu published an article in which he described administering doses of red or white phosphorous to one dog and a few sparrows. He considered that the deaths of the animals which had ingested the white phosphorous and the survival of those which had ingested the red phosphorous proved the harmfulness of the former and the innocuity of the latter. He did not take into account that the workers did not ingest either substance.[5]

Only under the Third Republic did the French medical profession modernize. This was less a result of the efforts of doctors than a part of the ideological struggle of the Republic vs. the Church, of science vs. superstition, of modern urban university-trained professionals vs. traditional rural healers, including the Catholic sisters who visited the poor with herb tea and consolation. New medical faculties were opened at Nancy in 1872, at Lille in 1875, at Lyon and Bordeaux in 1873-1874, and at Toulouse, Marseille, Nantes, and Rennes in the 1890s. The faculty in Paris was greatly expanded and improved with the latest in chrome and nickel operating room facilities. Because it was easier for provincial petty bourgeois to enter one of these new schools, especially after the requirement of a classical *collège* education was dropped, the training of *officiers de santé* ended by law on November 30, 1892.[6] To ensure the cooperation of doctors with the 1893 law on free medical aid to the poor, Parliament wrote into the 1892 law the right of doctors to unionize, but only in defense of their fees to individuals, thereby making their fees to the State, to *départements*, and to communes not the subject of bargaining. Shortly after, by a decree of July 1893, the scientific requirements for a medical degree were raised in physics, chemistry, physiology, histology, anatomy, microbiology, and parasitology. The hospital residency was also lengthened. Thus, for the first time, the French medical student of the 1890s could be sure of learning skills which would distinguish him from unlicensed practitioners. His social standing and political influence were to grow in consequence.

However, the period of this study was the period of emerging professionalism and dignity for French doctors. As we shall see, one doctor in particular tried to use the white phosphorous controversy to enhance his own medical specialization and the public reputation of the entire profession. Most of the contacts of the medical profession with the match workers were marked by the ignorance, lack of systematic investigation, and unwillingness to take a militant position on industrial health which was usual within the medical profession.

In 1845, twelve years after the first matches were manufactured in Wurtemburg, the first cases of necrosis were observed there. In that year, no fewer than seven doctors within the French Academy of Medicine were discussing whether necrosis attacked through the teeth or through the gums, a debate still unsettled 50 years later. By the 1850s the first five French cases were reported in Paris and the medical profession made an effort to develop some influence on occupational health. The *Conseil de Salubrité de la Seine* developed suggestions from the *Conseils de Salubrité* of the Nord, the Gironde, the Rhône, and the Bouches-du-Rhône into rules for the handling of white phosphorous. Manufacturers were to separate workrooms, provide ventilation and install updraft chimneys over the chemical basins. The Seine Council called for a ban and added the suggestion that only workers with healthy teeth be hired. Although doctors had not proved that necrosis attacked workers with damaged mouths, they asked for regular medical interference with the capitalist's right to dispose of his labor force. There is no evidence that they received that power in any systematic way.

By 1856, the year of the invention of the Swedish or safety match, Dr. Ambroise Tardieu wrote a lengthy summary for the *comité consultatif d'hygiène publique* in Paris. There had been an additional dozen cases in Lyon in the previous nine years. He concluded that even in those factories where the exposure had been diminished by the means described above, nothing but the substitution of the new red phosphorous would put an end to the problem. He mentioned that five of the largest manufacturers had petitioned His Excellency the Minister of Agriculture, Commerce, and Public Works to enforce the substitution of red phosphorous for white. Their businesses were suffering from fires, which were more frequent with white phosphorous, but they did not want to give an advantage to small manufacturers. The owner of the only white phosphorous factory in France also promised all cooperation if the government insisted on the change. The French holder of the patent on red phosphorous was willing to place it in the public domain.[7] Why then did the Minister of Commerce ignore Dr. Tardieu's report and the willingness of all concerned?

The answer was that France was in competition for the building of empire and the capture of foreign markets. According to Dr. Tardieu, France's serious competitor on the international market was England, which produced better quality but more expensive matches. Germany produced cheaper matches than France but had not yet entered the Australian and South American markets. Tardieu assured the government that

"...there the market is so securely French that a slight increase in price could not seriously hurt the export situation."[8]

Ironically, although France continued to produce white phosphorous matches, by the 1890s France had lost its export markets to Japan, Sweden, Germany, and Austria.[9]

Next, a study published in 1861 by Dr. Ambroise Chevallier introduced a new complaint. He wrote, on the authority of personal correspondence with an unnamed priest, that if a woman wanted to abort, she deliberately took a job at a match works.

"Any pregnant woman who performs a certain job in the chemical match factories miscarries, or if she does not miscarry, the child which she brings into the world is puny, weak and does not survive. I have however seen some of them vegetate for a month or two, but that is all. These accidents are constant and common to all the women who handle the paste which is applied to the match sticks..."[10]

For this reason he suggested urgent government intervention to eliminate women from the workforce. For male match workers he proposed the usual hygiene measures and the exclusive use of red phosphorous.

Although his suggestions were also ignored, the warning on spontaneous abortions, also known as miscarriages, was accepted as common knowledge by the doctors who cited Chevallier for the next 36 years. Apparently, no other doctor collected evidence on this subject until 1897.

In 1897, in the aftermath of the Pantin-Aubervilliers sickleave rate of 36%, there was a renewed interest in serious socio-medical investigation. Dr. Francois Arnaud, with seven years of practice at the Marseille match factory, wrote the most complete study on the subject. Specifically addressing the maternity issue, he compared the fertility of 350 women match workers to 120 women match box makers in Marseille over a three-year period. For every year of the study, the match workers' rate of live births was higher and their rate of stillbirths was lower. He hoped that this

> "simple statement of fact would suffice to put an end to the legend that has taken root in the scientific community on the subject of the tendency of women match workers to miscarry."[11]

Still he concluded that only a ban would put an end to the more general problem of necrosis.

It should be borne in mind that public health doctors rarely kept such longterm records of such a large sample of workers on any question of industrial health. Rare was the doctor who sustained an interest in a chronic condition when medical science was hard pressed to show results in the treatment of acute conditions. However, one doctor did distinguish himself in these years by his zeal in making himself an expert on necrosis and an ally of the match workers: Dr. Emile Magitot.

Emile Magitot, a disciple of Claude Bernard and de Broca, wrote his medical thesis in 1857 on the development and structure of human teeth and began operating on necrosis victims in 1874. In that year he wrote a report on his work for the *Société de Chirurgie de Paris*.[12] In the same year he was named to the Academy of Medicine and immediately started an anti-necrosis campaign among his fellow doctors, securing votes in favor of a ban from the *conseil d'hygiène de la Seine*, the *comité consultatif d'hygiène* and the Academy of Medicine. As we know, shortly thereafter at the end of 1889, there was an unsuccessful attempt to outlaw white phosphorous in the Chamber of Deputies but the Ministry of Finances remained unmoved.

After the match workers had captured public attention with their strike of 1893, Emile Magitot made himself known to them. By the 1894 Second Congress of the Match Workers' Federation, Magitot had advised them that improving hygiene was the answer and that all the medical authorities were already in favor of a ban. An article of his to this effect was published as part of the Federation's brochure of the 1894 Congress. From that time provincial match workers wanted only Dr. Magitot to treat them; the Marseille local secured 150 francs from the Marseille City Council for a woman worker named Rosine Cayol to travel to Paris.[13] Next, the Federation leadership asked the Ministry of Finances to transport all necrosis sufferers to Paris for the same reason, much to the Administration's annoyance. Magitot initiated the Academy of Medicine's renewed resolution of March 1895 in favor of a government ban and another resolution in favor of the appointment of a research committee. This committee was to seek the means of minimizing damages until such time as the ban might take effect. Unfortunately, there is no evidence that this committee was appointed. Emile Magitot worked on the problem alone for the next two years to the indifference of the government and the growing antagonism of his colleagues in the Academy.

When in October 1896 the Ministry of Finances did call for a medical committee, it was in the heat of the financial and production crisis which had been provoked by the high rate of sickleave at Pantin-Aubervilliers. Magitot was a member of that committee but he was not its chair. The chair was Dr. Emile Vallin, the editor of the *Revue d'hygiène publique et de politique sanitaire*, who had not distinguished himself by a particular interest in necrosis. In January 1897, Dr. Vallin published the committee's findings and recommendations for the future: the ban as soon as possible; in the meantime, closed machinery, better ventilation, a lower percentage of phosphorous in the chemical paste, a cooler temperature to the paste, the rotation of workers from more exposed to less exposed jobs, and the careful selection of only workers with sound teeth. None of these suggestions was new.

Dr. Magitot was unsatisfied with merely filing the report. He wanted to press the doctors' moral and scientific advantage to the point of results. He aired his views beyond the medical press: "There is no successor to white phosphorous,"[14] he wrote in the *Revue des Deux Mondes* in January 1897. The minutes of the Academy of Medicine sessions in the months of February and March 1897 indicate that the differences between Magitot's view and the other doctors' views had widened into a split.[15] Magitot had always stressed the importance of hygiene over the proposed ban, but he now declared himself to be in favor of hygiene alone. He claimed first of all that the ban would be long in coming. He urged his fellow doctors to seize this opportunity to gain public and governmental respect by resolving the crisis. Magitot became increasingly shrill in his insistence that this victory go not to industrial engineers for a new chemical compound but to doctors and to himself in particular.

He published a particularly excited polemic on March 1, 1897 in the *Revue des Deux Mondes.* In this article he stressed the importance of demineralization as a factor in the development of necrosis. Therefore, he wanted the continued distribution of free milk to restore the lost minerals. Furthermore, he suggested installing individual ventilation over each worker's site, dipping matches inside a box that would protect workers from the fumes, installing equipment which would measure air movement and would react to excessive phosphorous fumes. By these means plus the immediate removal of any worker who developed a case of necrosis, he claimed, in the most strident language, that the problem would be resolved.

"This is how the hygienist, armed with these weapons, can expect to immediately render any factory whatsoever healthy". (underlining supplied)[16]

Magitot pointed to the factories at Aix and Algiers which had excellent ventilation and no necrosis cases. However, by "any factory whatsoever", he was really referring to the Pantin-Aubervilliers factories.

Some of the other doctors on the committee believed that those buildings were built so that adequate ventilation was impossible. To the charge that the State had introduced some of these hygiene measures without satisfactory results, Magitot answered that the State's efforts had been merely empirical and "absolutely illusory...worthless". Referring to the reception that his ideas had received at the Academy of Medicine, he wrote:

"They don't want to hear about sanitation. There is for all of them only one solution and any voice that dares to plead the cause of hygiene is not heeded at all."[17]

He lamented that even the workers now stood in the way of his proposals:

"Minister (of Finances) Doumer... visited the factories of

125

Paris...He announced their closing and the awarding of all sorts of assistance to the workers...a strike was imminent. Therefore, all radical measures are postponed. They are postponed today, they will be postponed tomorrow, it's an impasse."[18]

His most radical suggestion was the wholesale relocation of the match industry in pursuit of perfect teeth:

"A factory in a Celtic area such as the Auvergne, for example, or in a black African area, would be ideal."[19]

The Match Workers' Federation was, of course, particularly incensed at this suggestion from their erstwhile bourgeois ally. Union officer Jacques Aschbacher denounced him at a public meeting:

"This false friend of the match workers has passed over to the enemies of the workers."[20]

Magitot finished this article in the *Revue des Deux Mondes* by claiming that anyone who did not agree was choosing barbarism over science. His opponents in the Academy of Medicine were, he claimed, guilty of abdicating their social responsibility as scientists and turning their backs on an assured victory for modernity.

On the day after the publication of this article, there was a meeting at the Academy of Medicine. The other doctors struck back. To Magitot's claim that the necrosis-free Aix and Algiers factories provided proof that hygiene was effective, Dr. Vallin snapped that not all factories could work all year round with their windows open. He might have added that Algiers was separately administered and that both factories were new: Aix had opened in 1894, Algiers in 1885. Furthermore, Vallin added that everyone knew that Arabs had perfect teeth; therefore, it was inappropriate to compare the Algiers factory to French factories. No one mentioned Marseille where the climate was also good enough to work with the windows open but there had been several cases of necrosis.[21] Next, Magitot read a flattering letter which he had received from the Director of an unnamed match factory. Dr. Roussel told the members of the Academy that the Director in question was hardly unbiased because he was a relative of Magitot's:

"These are family compliments which, in our opinion, have no place before the tribune of the Academy."[22]

Dr. Vallin was less interested in Magitot's motivation or in gaining prestige for the medical profession than in solving the problem quickly. With pique he read into the record a long complaint against

Magitot who, he claimed, wanted to interfere with the production of matches by "incessant" medical care and "almost daily" dental exams and urinalyses. Magitot, in his opinion, was wielding phosphorism and necrosis as a sword of Damocles over the heads of the factory directors. Furthermore, Magitot was responsible for encouraging the workers to make excessive wage demands in compensation for risk; to "raise the specter" of phosphorism at the slightest sign of ill health; to insist on months of paid sick leave, liters of milk, and trips to the country.

What was Emile Magitot's motivation? Was he a friend of labor? Why was he, alone of all the doctors in the Academy of Medicine, so eager to make the victory over white phosphorous a doctors' victory? Dr. E. Sauvez, the author of his obituary, shed light on the character and motivations of this unusual doctor. Sauvez explained that at the time when Magitot chose to follow in his father's footsteps as a specialist in diseases of the mouth, dentists and even oral surgeons were considered inferior to other doctors.

"It was perfectly understood that a surgeon who specialized in the rectum or the nasal cavities had chosen a more respectable field than (his counterpart in) teeth!"[23]

Therefore, Magitot never deigned to perform the functions of a dentist: he eschewed the drilling and filling of caries in favor of surgery. He created the specialization called stomatology, founded the Society of Stomatology in 1888, and was its president until his death.

"Stomatology was a religion. He was its founder and its high priest."[24]

Magitot was known for his arrogance towards dentists; even after 1892 when the law raised the educational requirements for the practice of medicine and created a special diploma for dental surgeons, he refused professional recognition to anyone who had not taken a complete course of medical study.

"There are approximately 2000 dentists in France of whom sixty, perhaps one hundred, are medical doctors. Magitot wanted nothing to do with any but the latter; as for the 1,900 others,... he refused them the right to exist. Every time one of them approached him, whether it concerned a conference or any other matter, he answered with the most definite refusal..."[25]

One can well imagine how little he thought of the dentists who had been pulling the teeth of match workers and how eager he must have been, for strictly professional reasons, to turn their confusion and failure into a victory for his new branch of medical science.

When the majority of the Academy of Medicine rejected Magitot's opinion, they seemed to be reacting to his suggestion that they enter into a new relationship with workers, that they practice not only the treatment of illness but preventive medicine. Because the Match Workers' Federation had made phosphorous necrosis an organizing tool, such intervention could only have been seen by the public and the government as fighting on the side of the workers. In reality, the Academy of Medicine may have been reacting to a power play by an arrogant and power-hungry member who had not had the tactical political sense to court his allies and to be sure that they attended.[26]

There the discussion ended with Dr. Magitot clearly outvoted. He died of asthma at the age of 63 shortly afterwards. No other doctor cared to press the issue. From March 1897 to October 1898 the problem was strictly in the hands of chemists and industrial engineers. Ironically, the Ministry of Finances did implement some of his suggestions by further improving ventilation and keeping careful medical records on the teeth of the workers for several years after the substitution of phosphorous sesquisulfide. However, Magitot was not to receive credit for his work from his fellow doctors because he had made himself too great an irritant. Neither was he honored by the Match Workers' Federation after they had discovered that he was not motivated by a desire to be a champion of labor but to be the vanquisher of necrosis.

The medical profession as a whole unconsciously advanced the match workers' struggle in several ways from the latter half of the nineteenth century. By their excessive attachment to theory over practice, to reciting precedent rather than keeping accurate records on the match workers' health, they perpetuated the myth of reproductive hazards. By his fear of necrosis, the dental surgeon at Pantin-Aubervilliers authorized so much sickleave in 1896 that he alone caused the Ministry of Finances almost as much loss as the six-week strike of the previous year. Finally, Dr. Magitot, in his effort to advance the area of specialization which he had created, called even more attention to the problem. Yet it was the resolve of the match workers themselves never to relax their pressure which was ultimately responsible for their victory.

1. APP Seine, *Rapport sur les travaux du Conseil de Salubrité de la Seine 1849-1861*, 63.
2. Free of all restrictions including educational requirements by the law of March 2, 1792, the French medical profession was unable to provide enough men of proven medical competence for the armed forces. Therefore, by the law of 14 Frimaire Year III or December 4, 1794, only doctors of medicine or surgery were fully permitted to practice medicine. Schools were founded in Paris, Montpellier and Strasbourg to give a shorter course of study to the new category of *officiers de santé* who were needed at the front or in hospitals. The title was also bestowed to regularize the situations of a variety of medical practitioners who had lesser degrees but years of practice. In theory, after 1794 a candidate needed either three years of medical school, five years of hospital residency, or six years of apprenticeship to a licensed medical doctor. He passed a perfunctory exam by a *départemental* jury or, after 1854, by a medical school or university medical faculty.
3. Louis-René Villermé, *Tableau de l'état physique et moral*, ibid.; Ambroise Tardieu, *Dictionnaire d'hygiène publique*, Volumes I-III, Paris: J.B. Baillière et fils, 1852; *Etude historique et médico-légale sur la fabrication et l'emploi des allumettes chimiques*, Paris: J.B. Baillière, 1856; Theophile Roussel, "Sur les maladies des ouvriers employés dans les fabriques d'allumettes chimiques et sur les mésures hygieniques et administratives nécessaires pour assainir cette industrie", *Comptes rendus hebdomadaires des séances de l'Académie des sciences*, Paris: Bachelier, 1846, 292-295; *Nouveau manuel pour la fabrication des allumettes chimiques*, Paris: Manuels-Roret, 1847. Roussel became a deputy and a senator from the Lozère. He was instrumental in the passage of the law of 1873 against drunkenness, the 1874 law on wet-nursing, the 1889 law on abused and abandoned children, and the 1893 law on free medical care for the poor.
4. Haroun Jamous, *Sociologie de la décision*. Paris: Editions du Centre National de la Recherche Scientifique, 1969, 64-65.
5. Ambroise Tardieu, *Etude historique*, ibid., 42.
6. Jacques Léonard, "Le corps médical au début de la IIIe. République", *Third Republic/Troisième République*, Spring 1978, 107.
7. Ambroise Tardieu, *Etude historique*, ibid., 50-52.

8. ibid., 49.

9. T.E. Thorpe, Thomas Oliver, George Cunningham, *Use of phosphorous in the manufacture of lucifer matches. Reports to the Secretary of State*, London: Eyre and Spottiswoode, 1899, 493, 502.

10. Ambroise Chevallier cited in Francois Arnaud, *Le phosphore et le phosphorisme professionnel*, Paris: J.B. Baillière, 1897, 173.

11. Francois Arnaud, ibid., 238. His figures on births and miscarriages did not still the belief that white phosphorous provoked reproductive difficulties. His colleagues at the Academy of Medicine remarked that a study done in Marseille where ventilation was a simple matter of opening the windows all year round was not applicable to the more northerly factories.

12. Emile Magitot, *De la nécrose phosphorée. Rapport lu à la Société de Chirurgie de Paris*, Paris: n.p., 1874.

13. Marseille, Conseil Municipal, *Délibérations, 1895*, Marseille: Moullot, 1898, September 14, 1895, 224.

14. Emile Magitot, "Des industries insalubres", *Revue des deux mondes*, January 1, 1897.

15. Drs. Magitot, Roussel, Bouchardot, Lereboullet, Laborde, Vallin, "Sur l'assainissement de la fabrication des allumettes", *Bulletin de l'Académie de médecine*, meetings of February 23, March 2, March 9, 1897 in *Revue d'hygiène publique*, March 1897, 250-263.

16. Emile Magitot, "Des industries insalubres", ibid., March 1, 1897, 167.

17. ibid., 152.

18. ibid., 155. Although we know that in 1896 the Federation was far from calling another strike and that the Administration was far from offering all sorts of aid, it is worthwhile to know that in some bourgeois circles the threat of another match workers' strike had been taken seriously.

19. ibid., 166.

20. Jacques Aschbacher at a union meeting of March 14, 1897 cited in APP 1.408 bis.

21. In addition to Rosine Cayol, who was mentioned as a necrosis sufferer in 1895, when the English doctor Thomas Oliver visited Marseille in 1898 he found 22 or 23 workers on sick leave including one, aged 23, who had been ill for almost two years. He also found two workers who had recovered from necrosis 20 to 30 years earlier but whose mouths were badly damaged; one had an enlarged

gland behind her jaw and a piece of exposed maxillary bone. Both were working filling frames with clean match sticks. The French doctor in charge of the Marseille factory admitted to Dr. Oliver that there had been three necrosis sufferers since 1890 and 13 cases of incipient necrosis who had been immediately removed from the factory. An interesting comment from Dr. Cunningham suggests that none of the doctors on the Academy of Medicine's committee was really in a position to know how many cases had occurred at each factory; he complained that there were few figures available on the medical service at Aix, adding: "Some of the results I obtained seem very instructive as to the comparative worthlessness of statistics without some standard of comparison and a knowledge as to the difference which exists in the dental service and the calls upon it in the several factories...The returns were so differently compiled (at Marseille and at Aix) that I found it impossible to make any table for comparison with the Aix factory." Dr. Cunningham also reported that there had been one case at Algiers. Thorpe, Oliver, and Cunningham, ibid., 565, 567, 696, 697, 714.

22. Théodore Roussel at a meeting of the Academy of Medicine, March 2, 1897 in *Revue d'hygiène publique et de politique sanitaire*, Paris, March 1897.

23. Dr. E. Sauvez, "Dr. E. Magitot", *L'Odontologie*, XIV, May 1897, 296-297.

24. Dr. E. Sauvez, idem.

25. Dr. E. Sauvez, idem.

26. In an article which appeared seven months later a doctor named Riche, whose name had not appeared previously in the controversy, wrote that he would have voted with Magitot and Bouchardot had he been at that meeting of the Academy. He then displayed his medical ignorance by adding that he was not persuaded that necrosis was a disease distinct from generally bad teeth and gums. Furthermore, he believed that those cases of extractions which were followed by a worsening of the condition could be ascribed to poor antisepsis rather than white phosphorous. Dr. Riche, "La Question des Allumettes: le phosphorisme", *Journal de pharmacie et de chimie*, September 15, 1897, 244, 289.

CHAPTER VII: WOMEN AND THE WORKING CLASS IN PUBLIC OPINION: THE MATERNITY QUESTION

"This poisoning, because of the lamentable sight of the unfortunate workers stricken by phosphorous necrosis or the chemical sickness, made a much greater impression on the (public) imagination than the equally noxious but less noticeable mercury and lead poisonings."[1]

Indeed, it was the public image of Frenchwomen with deformed faces and sterile wombs that intensified the movement for a ban on white phosphorous. Match manufacturing had always been a majority-female occupation, and, as we know, the hazards had been common knowledge for 50 years. However, hazards which had left the French indifferent suddenly attracted public sympathy and government intervention in the 1890s because the French began to dismay over their falling birthrate. Consequently, the belief that white phosphorous was responsible for miscarriages and stillbirths created the necessary extra impetus for the ban. Although other workers dealt with toxic chemicals such as lead and mercury, which were also believed to be abortifacients, the women workers who were exposed to those chemicals were few and far between; for example, the primary sufferers from lead poisoning were housepainters, all of whom were male. Also, although other workers dealt with possible abortifacients, their employers were not the Ministry of Finances. This Ministry, as part of a republican government which claimed to be responsible to its citizens and responsible for the future greatness of France, was subject to public pressure to protect the health of the next generation as no private industry could be.

In Parliament, the issue of the infertility which was allegedly provoked by white phosphorous was raised several times. The first discussion occurred in 1889 as part of the debate over direct administration. Deputy Paul Peytral (Bouches-du-Rhône) suggested that only adult males be employed in match factories, a suggestion which fell on deaf ears in the end-of-the-year rush to decide on private versus public ownership.[2]

During the 1895 strike, sympathetic speeches highlighting the maternity issue were made on the floor of the Chamber. They were influential in securing the 500,000 franc special allocation. On March 15 of that year, the fourth day of Pantin-Aubervillier's strike, Deputies Emile Goussot (Seine), Maximilien Carnaud (Bouches-du-Rhône), and Antide Boyer (Bouches-du-Rhône) proposed an

132

amendment to the budget of the Ministry of Finances: the addition of 100,000 francs for higher wages and disability payments for the match workers. In support of the amendment, Goussot read into the record Dr. Magitot's claim that all match workers suffered from phosphorism, which, according to Goussot, was producing future generations of anemic paper soldiers and inadequate workers. Claiming that he spoke neither as a representative of a special interest group nor a political party, he challenged the Chamber of Deputies on humanitarian grounds to allocate the 500,000 francs. At this point, Minister of Finances Maurice Rouvier asked for the smaller sum of 50,000 francs in prize money for the invention of a non-toxic strike-anywhere match. When the latter request was granted without opposition, Goussot withdrew his amendment.[3]

However, the match workers kept the maternity issue in the public mind with their statements to the press. For example, speaking to the feminist journalist Eugénie Potonié-Pierre three months later, one woman match worker claimed:

> "Our little children sometimes live; we manage to raise them to the age of three with many a care and precaution...Afterwards, they die; they are so fragile, you know, that it's a miracle they live that long!..."[4]

When the match workers believed that a substitute formula had been found in April 1898, they published in their one year old union newspaper:

> "From now on there will be tranquillity in our homes and employment assured to many mothers who thought they could never return to work."[5]

By insisting on the maternity question, the match workers were attaching their particular cause to the national fear of depopulation. As the French birthrate fell more noticeably after 1875, the theme of population grew in proportion. (See Graph IV: French Fertility Rate 1846-1913). By expressing an interest in the production of healthy children by the working class in general and the match workers in particular, the members of Parliament were reflecting the national consensus that the defeat of 1870 had been due, in part, to the numerical superiority of the Germans. Therefore, in the last third of the century, Parliament was increasingly willing to challenge the prevailing economic and philosophic liberalism in favor of the health of working class women and children.

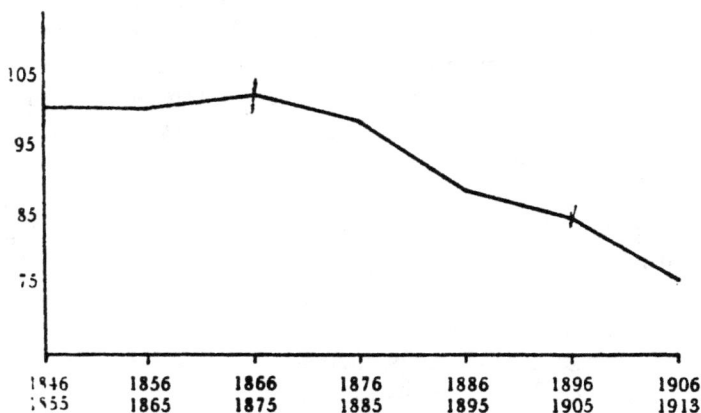

GRAPH IV: French Fertility Rate 1846-1913

To improve the well-being of children, Parliament passed several laws between 1879 and 1893 concerning the health conditions in public schools, such as the prevention of contagious diseases by compulsory vaccinations. As we know, at the workplace, children's hours were limited by the law of May 19, 1874; women's and children's hours were limited by the law of November 2, 1892. We must realize that public interest in the fertility of the match workers was only one manifestation of this national anxiety about the survival of French children.

The anxiety crystallized around the question of the bearing and raising of the next healthy generation of citizens in the same years as the match workers' struggle. It was popularized and polarized by two groups of population theorists: the neo-Malthusians and the repopulationists, those in favor of limiting births and those who were alarmed at the falling French birthrate.

The neo-Malthusians took their name from Thomas Malthus, who suggested in his 1798 publication, An Essay on the Principle of Population, that the alternatives to the problem of a population which grew more quickly than the food supply were either sexual restraint or famine, disease, and war. By the mid-nineteenth century,

British and American population theorists were using the term neo-Malthusianism to describe not sexual restraint but the use of birth control methods and devices.[6] In 1877, two British free-speech activists, Charles Bradlaugh and Annie Besant, attracted enough sympathizers to form an organization called the Malthusian League.[7] Two years later, the League began publishing a newspaper named *The Malthusian* and found sympathetic echoes on the Continent.[8]

The repopulationist movement grew in reaction. Only in the 1890s did the declining French birthrate come to the attention of several doctors and social demographers who prophesized the decline of French world power (and of the presumed universality of the French language). They analyzed a grab bag of causes and suggested legal remedies.[9] Moralists too took up the cry, pointing specifically to what they believed were changing sexual mores.[10]

In literature, the best example of the theme is in Emile Zola's novel *Fécondité* (1899) in which he contrasts the Froment and Beauchêne families. The large, loving, happy family of the engineer-turned-farmer Froment (symbolically named after a variety of wheat) prospers. The one sickly son of the socially ambitious, adulterous factory owner Beauchêne dies. As a final irony, one of the Froment sons takes over the Beauchêne's business. Marianne Froment, mother of 12, expresses the opinion that indeed her son's success is symbolic of the future: her kind will take over the world.

> "Expansion was inevitable; the earth was reserved for, possessed by the most numerous race....Since they had the numbers, they would be the strongest. The world would belong to them."[11]

In addition, statements like the following, by the fictional Dr. Boutan, Zola's mouthpiece for the repopulationist scientific community, were heard in bourgeois circles:

> "What was necessary were general measures, laws to save the nation: women assisted and protected from the first difficult days of pregnancy, exempted from heavy work, made sacred; later delivered in calm, even in secrecy if she desired, with no demand on her other than motherhood. Then the woman and child should be cared for, aided during the recovery and the long months of nursing (Zola favored 15 months of nursing) until the day when the child was finally in the world so that the woman could again be a healthy and vigorous wife. There were a series of precautions to take,

homes to create, pregnancy hostels, secret maternity wards, convalescent homes, in addition to laws of protection and aid for nursing mothers...In our democracy, woman, as soon as she is pregnant, should become sacred."[12]

In this atmosphere, Pope Leo XIII's encyclical of May 15, 1891, *Rerum novarum*, on Christian charity, made it fashionable for bourgeois ladies to serve in organizations like *L'oeuvre des enfants assistés* which distributed infant clothes and cradles to working class single mothers. This in-kind assistance was meant to persuade them to keep their babies rather than abandon them or send them to cheap wet nurses who were known for their high infant mortality rates.

In France, formal organization of the neo-Malthusian movement began during the same years as the match workers' agitation over white phosphorous. The first blow was struck in 1892 by Marie Huot, a nihilist *exaltée*, who gave a lecture in Paris in which she recommended *la grève des ventres* until revolution.[13] The following year, a Dr. Brennus published a book on contraception entitled *Amour et sécurité*. Both events created short-lived scandals in the press. Then in 1896, Paul Robin, an ex-member of the International, who was already known for his direction of an experimental coeducational school, launched the French branch of the Malthusian League, which he named the *Ligue de la Régénération Humaine*. The organization, which consisted only of Robin and a few friends for ten years, performed propaganda by word and deed: distributing literature, making speeches, and selling contraceptives. The League members also referred people to doctors who would supply other means. After their translation of a Dutch brochure on the latex pessary, *Moyens d'éviter les grandes familles*, in the fall of 1896 the mass circulation weekly *l'Eclair* ran a series contrasting the ideas of Robin with those of the repopulationist demographer, Dr. Jacques Bertillon. This gave the League such publicity that its membership list swelled and legal action against its existence was threatened.[14] One year later, not many of the new members had renewed their dues. Insolvency continued until 1901 when two large anonymous donations arrived.[15] The League also suffered from personality problems until the temperamental 65-year old Robin found a young dependable collaborator named Eugène Humbert. From that happy collaboration starting in 1902, the journal *Régénération* appeared monthly and the movement grew.

Once the organization was firmly planted, it flourished in France because there had long been an inarticulate mass movement. The French of all classes were receptive to contraceptive advice and supplies because they were already practicing family limitation by traditional means. Although the birthrate started to decline in the mid-eighteenth century, after 1875 it plunged. The tendency was particularly noticeable among the bourgeoisie who often limited themselves to one male heir. Yet there was a decline in births among all classes except some peasants and peasant-artisans who could set their children to work between the ages of five and ten. The traditional means of family limitation were delayed marriage, *coitus interruptus*, periodic abstinence, and prolonged lactation. In addition, Frenchwomen relied on abortion as their own method for asserting control when male self-control failed.[16] Modern methods such as prophylactics, vaginal sponges, douching solutions, and suppositories became increasingly available at pharmacies and by mail in the last third of the century. By the 1890s, a few Parisian doctors were even fitting the pessaries which the Dutch neo-Malthusian, Dr. Messinga, had recently invented.[17]

Acceptance of the practice did not, however, indicate acceptance of the theory. According to the French neo-Malthusians, overpopulation was an important cause, if not the primary cause, of the inequalities between the owning and working classes. If workers limited their families as did the bourgeoisie, the neo-Malthusians expected that there would be less competition among the next generation of workers. Higher wages and better living conditions would be the results. Most socialist leaders were indifferent or hostile to neo-Malthusian doctrine as another reformist diversion of the workers' attention from the immediate need to struggle. However, many anarchist and anarcho-syndicalist groups supported it and even sold contraceptives to raise funds.

In the 1890s, despite organizational difficulties, the expression of neo-Malthusian ideas and certainly the practice of neo-Malthusian methods continued to find favor in libertarian/syndicalist circles. Robin addressed *universités populaires* study groups, masonic lodges, and revolutionary groups as well as trade unions.[18] He received sympathetic coverage from the feminist newspapers *La Fronde* and *La Femme de l'Avenir*, the small circulation anarchist paper *Le Libertaire*, the large circulation socialist newspaper *La Petite République*, and even the mass circulation *Le Journal*. As greater success and the active membership of libertarian and feminist celebrities followed, branches of the neo-Malthusian League were founded in Paris, the suburbs, and the provinces. However, the theory was taking root in the fertile soil of a libertarian and syndicalist public which was eager to use it for the improvement of their and their children's lives, if not for revolution.

In the other camp, the bourgeoisie was slow to organize into formal associations, although bourgeois public opinion became repopulationist. (In private, the bourgeoisie continued to practice family planning.) They expressed particular concern about the working class' refusal to perform their duty to the future. In reaction, bourgeois critics provoked the trial of Dr. Brennus, author of *Amour et sécurité*, for pornography in 1895. (He was acquitted.) Occasionally a neo-Malthusian speaker was denied the use of a municipal hall. Public monies found their way into the *Oeuvre des enfants assistés*. There was talk of reestablishing the *tours*, infant-sized revolving doors in the walls of welfare offices. These doors had been designed so that one could abandon a baby anonymously but be sure it was safely inside. To facilitate marriages among workers, the marriage laws were simplified in 1896 and 1907, without effect on the falling birthrate.[19] Clearly, the Match Workers' Federation seized on a timely issue when they claimed publicly that the government's irresponsibility and procrastination were costing the French nation its future citizens. However, was their fertility in reality threatened? As we saw in Chapter III, census records and personnel dossiers indicate that the match workers of Pantin-Aubervilliers had a higher percentage of childless households than their neighbors or the French in general. The analysis of that data suggested that their childlessness was a choice related more to their age and marital status than to health considerations. If the match workers were visiting Paul Robin's twentieth *arrondissement* headquarters to buy contraceptives when they didn't want children, so were many of their working class neighbors.[20] The match workers were no more practicing *la grève des ventres* for political reasons than they were suffering from infertility due to their occupational hazard. However, in a majority female trade with an occupational health hazard, it was politically astute to agitate the specter of reproductive hazards during these years.

ENDNOTES CHAPTER VII

1. France, Ministère du Commerce, Office du Travail, *Poisons Industriels*, Paris: Berger-Levrault, 1901, 129.
2. Journal Officiel, Chambre, *Débats*, November 23, 1889, 131.
3. ibid., March 15, 1895, 943-945.
4. Eugénie Potonié-Pierre, "Les Allumettiers", *La Question Sociale*, June 1895.
5. *L'Echo des allumettiers*, April 1898.
6. Francis Ronsin, *La grève des ventres 19e-20e siècles*, Paris: Aubier Montaigne, 1980, 38.
7. ibid., 39. Bradlaugh and Besant provoked their arrest as pornographers for selling a birth control brochure, Dr. Charles Knowlton's *Fruits of Philosophy*. In acquitting them of the charge, British law acknowledged a distinction between pornography and scientific information.
8. The Dutch *Nieuw malthusieanse bond* was founded in 1881. A Stuttgart group, *Sozial Harmonischeverein*, limited itself to publication. Ronsin, ibid., 39 and 40.
9. Dr. Jacques Bertillon, the most prolific writer, attacked the neo-Malthusian theories, which he called criminal propaganda, with traditional free-market economic arguments. Overpopulation was not responsible for unemployment; the unemployed themselves were "inadequate workers who find no work because they are incapable of doing it well." Reducing the numbers of competitors for jobs would only reduce the demand for goods and services, he claimed. Reducing the number of French competitors for jobs would only result in the flow of jobs to neighboring countries with high birthrates or the flow of immigrants into France. He also blamed the democratic spirit as it had evolved from the French Revolution. This had weakened the influence of religious authorities over fertility. The Napoleonic Code, product of the Revolution, had also changed the inheritance laws so that families had to divide their property among their heirs. In addition, the desire of the common Frenchman and Frenchwoman to rise socially had led to an excessive desire to save money; consequently, society as a whole no longer respected people who struggled to support a brood. Bertillon also pointed to some material obstacles to repopulation such as complicated marriage formalities and the urban housing shortage, both of which contributed to late child-bearing by Frenchwomen.

His recommendation was State intervention on behalf of large families: special political rights for the fathers of three or more children, public housing, bonuses upon the births of all children after the third, and organized festivals to honor children. Admirers of Vichy France will recognize the origin of that regime's family policies. Dr. Jacques Bertillon, *De la dépopulation de la France et des remèdes à y apporter*, Paris: Alcan, 1896. Arsène Dumont's stated goal was to encourage reproduction among the upper classes. For this reason, he reviewed the legislative attempts to encourage fertility since 1883. He found these efforts timid and insufficient. Changes in the inheritance tax had been too slight. Lower taxes for families of seven or more children, he felt, would only be an encouragement to those who already had six. Ironically, in opposition to Bertillon, Dumont placed his greatest hope in suppressing Church influence. He considered the teaching nuns to be role models of sterile virginity to too many French girls. His solution was to close all the convents which were not specifically authorized and to open more dancehalls. Arsène Dumont, *Dépopulation et civilisation*, Paris: Vigat, 1890. Finally, Emile Levasseur's three-volume work offered the best statistical analysis possible with the data of the 1890s. He measured the French in every way: age at marriage, average number of children per marriage in each *département*, annual numbers of stillbirths and infanticides, and the ratios of each to live births. Yet, he was unable to explain why the French were the only European people to experience such a decline. Emile Levasseur, *La population francaise*, Paris: Rousseau, 1891.

10. Tommy Fallot, *Communication sur l'organisation de la lutte contre la pornographie faite au Congrès de l'association protestante pour l'étude pratique des questions sociales*, Marseille: 1891; Gaufres, *La corruption de la jeunesse par la presse pornographique*, Saint-Etienne: 1897.

11. Emile Zola, *Fécondité*, Paris: Charpentier, 1915, 481.

12. ibid., 275-276. In 1913, on the eve of the war for which the French had so worried about providing soldiers, Parliament finally passed a law which subsidized four weeks of maternity leave.

13. Ronsin, ibid., 44. Marie Huot had already come to the attention of the public as the founder of the *Ligue populaire contre la vivisection* or Anti-Vivisection League for her umbrella attack on Dr. Brown-Séquard, who operated on live rabbits at the Collège

de France.

14. The League received 500 letters and the dues of 150 new members by January 1897. The threats of legal action were dropped only because of the importance of the Dutch author of the brochure and the Dutch League's founders. Ronsin, ibid., 55-56.

15. Raoul Félice, *Les Naissances en France*, Paris: Hachette et Cie., 1910, 283.

16. Frenchwomen provoked abortions with savine and ergot of rye, sometimes accompanied by hot baths and violent exercise. Angus McLaren *Sexuality and Social Order*, New York: Holmes and Meier, 1983, 144 and Ronsin, ibid., 39.

17. Ronsin, idem.

18. Sympathizers in the Sud-Ouest, the Nord-Pas-de-Calais, the Jura, and Provence organized speaking tours for the Parisian neo-Malthusians. Nelly Roussel and Alexandra Myrial, radical feminist celebrities and journalists for *La Fronde*, participated in the movement. Sebastien Faure, prolific anarchist speaker and writer, was converted to neo-Malthusianism by Humbert. Georges Yvetot, secretary of the *Fédération des Bourses du Travail*, threw himself into the debate on the neo-Malthusian side in 1904. The trade union congresses of lithographers, leather, food, and ceramics workers voted resolutions on neo-Malthusianism between 1904-1911. The bronze workers, metal engravers, *Union des syndicats de la Seine* and *Fédération des Bourses du Travail* invited Paul Robin to speak. The *Bourse du Travail* of St.Denis was the headquarters of the St.Denis neo-Malthusian groups and pronounced itself as a whole in favor of neo-Malthusianism in 1904. Ronsin, ibid., 59-61. However, most socialist leaders rejected neo-Malthusianism for three reasons. It was reformist. Second, it relied on individual self-help rather than collective struggle. In the future socialist society, they argued, there would be no need to restrain population growth. Finally and most important, it was immoral because it violated the natural purpose of sex: parenthood. On this last point, see Angus McLaren, "Sex and Socialism in Nineteenth Century France", *Journal of the History of Ideas*, 1976, 475-492. McLaren contends that French socialists reflected the views of bourgeois society: they were prudish about sex as pleasure. Perhaps they took this stand of moral purity to compensate for their social critiques. As far as socialist publications were concerned, other than occasional

favorable coverage in *La Petite République* and consistent support from *La Guerre Sociale*, neo-Malthusians could not count on them.

19. Of greater assistance in increasing the number of marriages was the law of 1884 which legalized divorce and therefore made possible a second, presumably happier, marriage. Although all demographers agreed that the majority of births occurred to married people, the post-1884 remarriages did not produce a noticeable increase in births. Marriage was not a sufficient condition to ensure parenthood. It was not until the first decade of the twentieth century that the repopulationists organized nationally and began to see success in the suppression of information, although they were still unsuccessful in raising the birthrate. In 1905, Senator René Bérenger (Drôme) capped his career as a moral guardian of children and adolescents against sexually-oriented public entertainment, prostitution, and neo-Malthusianism by calling together 3,000 representatives of organizations to found the *Fédération des sociétés contre la pornographie*. This Federation and *L'Alliance nationale pour l'accroissement de la population francaise*, which was founded in 1907 by Dr. Jacques Bertillon, fought neo-Malthusianism with the same weapon which their British colleagues had wielded unsuccessfully against Bradlaugh and Besant: the charge of pornography. Since the French law was silent on contraception, public speech and sales were perfectly legal. By 1910, however, courts began to condemn neo-Malthusians for pornography on the grounds that their literature or contraceptives were displayed where young people could see them or in a manner contrary to public decency (*contraire aux bonnes moeurs*).

20. The *Ligue de la Régénération Humaine* had headquarters at 27, rue de la Duée, in the 20th *arrondissement*. The match workers lived nearby in the 20th, la Villette, Pré St. Gervais, Pantin, and Aubervilliers.

CHAPTER VIII: WOMEN'S PARTICIPATION IN LABOR UNIONS: THE CASE OF THE MATCH WORKERS

The women match workers were model trade unionists at a time when their sisters in other factories and in the sweated trades remained unorganized or were unable to assure the survival of their unions.[1] The behavior of the match workers and their cohorts in the Tobacco Federation was so exceptional, in fact, that it merits closer analysis. In this chapter we will examine why these unions organized successfully and why the men in the Tobacco and Match Workers' Federations cooperated so heartily with their female fellow workers at a time when most union men were hostile to the presence of women in the workforce at all and more hostile to those women who sought to join their unions. Finally, we will look at one aspect in which the two Federations were different: the unequal representation of women among their leadership.

In all of French organized labor at the end of the century, the State manufactures were outstanding for their number of female union members. Although the majority of women who worked were in textiles and clothing, only 2,300 of them had been able to form and sustain unions by 1900.[2] In comparison, all 18 tobacco factories and all seven match factories were unionized and almost all of the 18,000 female tobacco workers and 1,400 female match workers belonged. Of the fifteen *départements* with the highest percentages of female union membership in 1900 (the first year for which statistics on female membership are available), six *départements* had only one union which included women: a tobacco local.[3] (See Chart XXI: Labor Unions in *Départements* with Greater than 15% Female Union Membership in 1900 and Chart XXII: Female Tobacco Worker and Non-Tobacco Worker Union Membership in the Fifteen *Départements* with the Highest Female Union Membership in 1900). Therefore we know that the tobacco and match women did not join merely because they performed factory labor; nor did they join because their particular *département* was favorable to unionism or to female participation in unions. Rather it was that female tobacco and match workers were able to form viable unions for reasons unrelated to their sex: political conditions in the Third Republic induced the State to tolerate unionization in those industries and services which it owned. There was some struggle where the work was deemed essential to the national defense, but even the railroad and arsenal workers were eventually permitted to unionize.[4]

CHART XXI: LABOR UNIONS IN *DÉPARTEMENTS* WITH GREATER THAN FIFTEEN PERCENT FEMALE UNION MEMBERSHIP IN 1900

Industry	Tobacco	Tobacco	Tobacco
Name of union	Syndicat des ouvriers et ouvrières aux tabacs	same	same
City	Châteauroux	Riom	Morlaix
Département	Indre	Puy-de-Dôme	Finistère
Number of women in union	1197	498	800
Number of union women in that industry in France	3181	same	same
Percentage of female union membership	96%	92%	92%
Percentage of *départemental* female union membership represented by that union	68%	38%	30%

Tobacco	Tobacco	Tobacco
Syndicat des ouvrières aux tabacs de la manufacture d'Orléans	Syndicat des ouvrières des manufactures de tabacs, section d'Orléans **	Syndicat des préposés à la manufacture de tabacs
Orléans	Orléans	Morlaix
Loiret	Loiret	Finistère
150	150	10
3181	3181	3181
100%	100%	30%
23%	23%	30.2%

** This is probably the same union as the one which preceeds it. The first one was founded in 1892, the second in 1898.

Tobacco	Tobacco	Tobacco
Syndicat des ouvriers et ourières aux tabacs	Syndicat des ouvrier et ouvrières aux tabacs	Syndicat non fédéré des ouvriers et ouvrières de la manufacture des tabacs
Tonneins	le Mans	Orléans
Lot-et-Garonne	Sarthe	Loiret
221	295	20
3181	3181	3181
87%	87%	66%
27.6%	27%*	23%

* Guilbert's figure is 24.4%. This does not take into account the *chambre syndicale des ouvriers et ouvrières en manches de parapluies:* 75 men, 30 women.

Linen	Umbrella handles	Noodles
Chambre syndicale des tisseurs de toile, linge de table et similaires	Chambre syndicale des ouvriers et ouvrières en manches de parapluie et ombrelles réunis	Chambre syndicale des ouvriers et ouvrières en pâtes alimentaires
Voiron	Vibraye	-
Isère	Sarthe	Isère
35	30	16
35	30	16
15%	29%	43%
23.3%	24.4%	23.3%

Mother-of-pearl	Military equipment	Textiles
Chambre syndicale des ouvriers nacriers	Syndicat de l'équipment militaire	Union syndicale de l'industrie textile et similaires
Vinay	Clermont-Ferrand	Vienne
Isère	Puy-de-Dôme	Isère
4	2	1100
4	2	1263
5%	10%	32%
23.3%	37.7%	23.3%

Textiles	Textiles	Cotton
Chambre syndicale ouvrière textile	Syndicat des ouvriers du textile	Chambre syndicale de l'industrie cotonnière
Bagnères-de-Bigorre	Lavelanet	-
Hautes-Pyrénées	Ariège	Mayenne
128	35	411
1263	1263	575
75%	21%	65%
18.3%	15%	42.6%

Cotton	Cotton	Spinning and Carding
Chambre syndicale des ouvriers de l'industrie cotonnière	Chambre syndicale des travailleurs de l'industrie cotonnière	Syndicat professionnel des ouvriers et ouvrières des usines de filatures et carderies
Flers	Condé-sur-Noireau	Tenay and Argis
Orne	Calvados	Ain
64	100	460
575	575	460
53%	40%	51%
19.5%	19.3%	37%

Combs	Metal	Rubber
Chambre syndicale des ouvriers et ouvrières en peignes et des parties s'y rattachant	Chambre syndicale des ouvriers sur metaux	Syndicat des employés, ouvriers et ouvrières des manufactures de caoutchouc
Ezy	la Vallée du Ralin	-
Eure	Haute-Saône	Puy-de-Dôme
78	60	49
92	60	49
29%	30%	31%
15.1%	16.5%	37.7%

Clerks	Glovemaking	Shoes
Union fraternelle des employés de commerce de la ville	Union syndicale des ouvrières gantières	Chambre syndicale des ouvriers et ouvrières en chaussures
Grenoble	Grenoble	Morestel
Isère	Isère	Isère
48	232	8
48	232	8
10%	100%	40%
23.3%	23.3%	23.3%

Shoes	Combs
Union syndicale des ouvriers et ouvrières de la cordonnerie	Union des ouvriers en peignes et similaires
Tonneins	Oyonnax
Lot-et-Garonne	Ain
96	14
104	92
51%	9%
27.6%	37%

Information from Guilbert, 35-6.

CHART XXII: FEMALE TOBACCO WORKER AND NON-TOBACCO WORKER UNION MEMBERS IN THE 15 *DÉPARTEMENTS* WITH THE HIGHEST FEMALE UNION MEMBERSHIP IN 1900

Département	Percentage Male Union Members All Trades	Percentage Female Non-Tobacco Union Members	Percentage Female Tobacco Union Members
Indre	32.3	-	67.7
Mayenne	57.4	42.6	-
Puy-de-Dôme	62.3	5.0	32.7
Ain	63.0	37.0	-
Finistère	69.8	-	30.2
Lot-et-Garonne	72.4	10.0	17.6
Sarthe	73.1	2.5	24.4
Isère	76.7	23.3	-
Loiret	77.0	-	23.0
Orne	80.5	19.5	-
Calvados	80.7	19.3	-
Hautes-Pyrénées	81.7	18.3	-
Haute-Saône	83.5	16.5	-
Eure	84.9	15.1	-
Ariège	85.0	15.0	-

Figures from Guilbert, 30-31.

In 1888 when the Marseille and Lyon tobacco locals had just formed, the Minister of Finances Paul Peytral, the Inspector of Finances, and the Director of the Lyon factory all examined their statutes for evidence of illegalities. Unlike management in private industry, these State managers were obliged to adhere to the letter of the law: the union statutes were legal; the union would be allowed. In 1889, the new Minister of Finances Maurice Rouvier reexamined the statutes and found the same.

On the local level, authorities were sometimes sympathetic in their attitudes towards unions. Even among those republicans who were not sympathetic towards institutions of class struggle, some held dear the principle of freedom of association. As early as 1883, when the Le Havre tobacco workers struck for five days, the city authorities so respected their rights that they refused to call out the police.[5] In 1887, when the Marseille tobacco women struck because of their foreman's severity and rudeness, members of the General (Departmental) Council and the City Council validated their grievances by attending their public meeting. After 1884, when labor unions were sanctioned by law, the organizing drive in the tobacco industry started in earnest. From 1887 to 1890, and during the united tobacco and match organizing drive from 1890 to 1892, whenever possible the local authorities and politicians were invited to attend the organizational meeting to stress the pacific and legal nature of the union. So it was at Le Mans in 1892 when the prefect of police presided over the organizational meeting of the tobacco local; at Tonneins, the mayor himself sat on the podium.

There is evidence that these unions succeeded precisely because the workers understood their special status so well. For example, they recognized that their bosses at the individual factories did not hold the ultimate power and were replaceable. Therefore, both in organizing on the local level and in bargaining on the national level, they looked to their elected officials and the makers of political and financial decisions in the Ministry of Finances. If the police and city council were on their side, they knew that they could agitate without significant reprisals on the local level. If, in addition, they could win the sympathy of their deputies, the parliamentary budget committee, and the Minister of Finances, what did it matter if their factory director at first refused their grievances?

But making strong unions required more than a government which was too legalistic and often too preoccupied with other matters to halt the ferment in these two small industries.[6] For that, it took labor militants.

Indeed, the French tobacco industry and later the match industry attracted an unusually militant type of woman. In temperament, women tobacco workers were believed to be different from other working women. They were known for their non-traditional behavior such as smoking in public and bursting into protest over work rules long before the formation of their first union. The women at the Toulouse tobacco factory had founded a mutual aid society, which may have provided strike funds, since 1852. Many accounts of their work suggested that the handling of tobacco produced this chronic state of nervous irritability, but modern day tobacco workers do not raise this complaint. It is therefore more likely that there was a certain self-selection among the women who sought work there after which, in the words of Charles Mannheim, the author of a comprehensive

150

1902 history of the two industries, "the apprentices were goaded on by the experienced women."[7]

From 1870 to 1887 there were 12 recorded work stoppages at nine of the 18 tobacco factories. Unlike other working women who rarely went on strike and usually did so only when wages were cut,[8] the tobacco women often struck for a variety of reasons: for the firing of a particularly resented worker, foreman, or director (Toulouse in 1870 and 1874, Nice in 1876, Tonneins and Lyon in 1876) or for the rehiring of a comrade (Lyon in 1882). In addition, they were known for refusing to work with materials which they considered to be poor in quality (Morlaix in 1874, Le Havre in 1883, Marseille in 1887) and for refusing to adopt new work methods (Dieppe and Toulouse in 1875).[9] However, most of their job actions between 1870 and 1887 failed. Why then did they continue to lash out in anger when women workers in other industries were so easily stymied?

The answer is that in comparison with the disregard shown by private employers, the State treated its workers so much more respectfully and, in case of strike, so much less violently that the tobacco women were confirmed in their sense of special status. For example, when the tobacco workers of Toulouse went on strike out of personal animosity against their director in 1870, he resigned. In most years work stoppages were met with the denials of the grievance(s) but no punishment worse than a few isolated firings or punitive layoffs for the alleged instigators. Never did the management of a tobacco factory fire the entire staff or make across-the-board paycuts as private industry did. In the meantime, deputies, senators, and even Undersecretaries of State were known to exert their influence on the tobacco workers' behalf, giving them the impression that the next time they just might win.[10]

Only once, in 1875, was there a strike in which outside force was used. When Toulouse struck for the third time in five years and some workers stayed out for six weeks, the army was called in. Mannheim does not, however, indicate that anyone was injured.[11]

An example of one of these lost strikes which nevertheless gave the women a sense of accomplishment occurred in 1883 in Le Havre. When the women closed their factory over the quality of tobacco, management fired the five ringleaders. On the next day, 250 women were on strike. Four days after the initial walkout, an Inspector with full powers to negotiate arrived from Paris, examined the issues, and refused any concessions; only then did the women return to work.

151

Although the five ringleaders had to sign a letter of respect and submission to the Minister of Finances, they were readmitted. In sum, the strikers lost the strike but won their dignity: they had eschewed the opinion of their factory director, received the sympathy of the police (whose inactivity received special mention from Mannheim), and made a representative from the center of power travel to them. This 1883 strike set precedent. The tobacco workers would never again consider a refusal from a factory director as definitive.

Four years later in Marseille the story was repeated but with even more lasting results. On January 5, 1887, the women began a strike over the quality of the tobacco they had to handle; in addition, they complained of a particularly offensive foreman who had almost been the cause of an earlier strike. Three days later, there was a public meeting in the presence of local officials and journalists. By then the Minister of Finances, Albert Dauphin, recognizing that his factory director had lost authority, sent an engineer and an inspector to inquire. To everyone's surprise, the women of Marseille insisted on a measure of workers' control: they wanted half of the seats on the committee of inquiry. This was refused. Then, 150 women testified in favor of the replacement of the foreman, who was indeed removed. (Mannheim provides no further information on the quality of the tobacco). When the women learned, however, that their ex-foreman was working elsewhere on the premises, they insisted that he be removed once and for all. On January 12, the Marseille union was born of this protest.[12]

Between 1888 and 1890, the trail is lost. By 1890, the tobacco works in Paris, Dijon, Nancy, Châteauroux, and Nice had organized and founded the *Fédération nationale des ouvriers et ouvrières des manufactures de tabacs de France*. In 1891 another six locals joined (Dieppe, Toulouse, Lille, Bordeaux, le Havre, and Morlaix); in 1892 Riom, Tonneins, le Mans and Orléans were added; by 1893 all 18 tobacco factories were represented. But the locus of power had shifted from those namelsss women on the shop floor, bursting into anger over unspecified insults and rudeness from foremen to Secretary-General Ducros[13] at the *Bourse du travail* in Paris.

The women were the backbone of the individual unions but men became the leaders of the Federation. Exactly how the power had shifted to a city whose tobacco local had not distinguished itself by early militance or whether the workers perceived any slippage of power is a question lost to history. Of the uncited, nameless women who built the tobacco workers' militant traditions through the 1870s and 1880s we have only a few lines on a Marseille woman named Marie Jay. She claimed to have argued successfully against the first president of the Marseille union, a woman who was opposed to joining the Federation.[14]

Furthermore, Jay and a co-worker named Marie Deleuil claimed to have been among the organizers of Riom and Orléans. However, since Riom and Orléans joined in 1892, at least five years after the formation of the local in Marseille, it is difficult to see Marseille as the "Motherlode" from which organizers spread throughout the country.[15] Certainly those vigorous tobacco workers from Marseille, Châteauroux, and Morlaix continued to be active locally and were delegates to Federation congresses. Also the description of job actions at the local level sounded the same as they had before the formation of the Tobacco Federation. For example, on February 9, 1895, a strike broke out at Dijon when the factory Director cast aspersions on the private life of a woman worker. The Chief Inspector was dispatched from Paris. Minister of Finances Alexandre Ribot promised an inquiry. Only then did work resume.[16] However, on the national level, job actions were now translated into the more formal language of official trade union, labor exchange, and *Office du travail* reports: date of strike, duration of strike, number of strikers... In the translation, the women's spirits were lost.

Once the match workers became directly responsible to the Ministry of Finances, they too assumed that they benefited from the traditional privileges of the tobacco workers. The fact that the match workers who had worked under the private company before 1890 were rehired as a block by the Ministry of Finances must have increased solidarity, the feeling that the workers were exactly the same but that only the boss had changed. The Ministry increased their sense of identification with their militant comrades by occasionally moving workers from matches to tobacco or vice versa as the need for labor changed or workers asked to transfer to another city. The Ministry further emphasized the unity of the two industries by counting the years spent in either industry towards the same pension. This was a situation in which the women match workers were bound to feel inclined to test the limits of their new status.

That is why the Match Federation seemed to emerge in 1892 fully grown with all of the experience which the tobacco workers had accumulated over the years. Unlike most new majority-female unions, the Match Workers' Federation was not born in the heat of a strike but was deliberately conceived and nurtured by the leadership of the Tobacco Federation, which provided it with an office, statutes, and an acting Secretary-General, Ducros.[17] Before they declared their first industry-wide strike, they had set their demands, elected their central committee in Paris, and empowered it to act for the whole.

However, there was another change after the formation of the two Federations. After their founding congresses, they spoke through the male voices of their elected union leaders. Leaving aside for the moment the question of why and to what extent these majority-female trades elected men to represent them, let us look at how the men and the women of the two unions approached the question which rent so many other mixed unions in the 1890s: the question of woman's proper place.

In other unions, when women overcame bourgeois society's hostility to join mixed unions or even all-female unions, they often met with the hostility of the male union members of that trade. The prevailing workers' philosophy on this subject was drawn from the anarchist and misogynist, Pierre-Joseph Proudhon, who argued that woman's place in society was either as *menagère ou courtisane*, housewife or harlot.[18] The argument was usually framed in the following terms. Woman's place was in the home where she cared for husband and children. It was assumed that women who worked outside the home were sexually promiscuous; another way of arguing this was to say that women needed to be protected from the whims of male bosses. An additional argument, which was common to workers regardless of political leanings, was that women displaced male heads of households by working for lower wages. The counter-argument, that there were also women heads of households and self-supporting single women, was made at the national congresses of those trades in which women were increasingly hired by employers regardless of the opposition of male workers. Therefore, rather than weaken an all-male union's power by refusing to organize this growing segment of the workforce, some unions decided tentatively and begrudgingly to allow women to join.[19]

This was the case for the *Société Générale des Chapeliers de France*, the Hatmakers' Union, in 1884. At that Congress several unaffiliated unions of women hatmakers were admitted on the condition that they were not to speak.[20] As another example, the workers of the *Fédération du Livre*, the principal typesetters' and printers' union, wrote into their founding statutes in 1881 their opposition to the employment of women typesetters. Even in those cases where the employer agreed to equal pay for men and women, the majority of unionized typesetters invoked the name of Proudhon as they opposed working with women for "moral" and "health" reasons. In later years, as employers continued to add women to their workforce, some male typesetters argued that the exclusion was weakening the Federation. At each union congress (1883, 1885, 1887, 1889, 1892, 1895, and 1900) the question was raised again. By a referendum of 1900, the typesetters voted to keep their union closed.[21] Finally at the 1910 Congress, the *Fédération du Livre* moved to accept female membership for a trial period of two years during which time the union would try to raise the women's wages to the level of the men's.[22]

In comparison, the Tobacco and Match Federations were unusual because there was no opposition whatsoever from union men to the employment or unionization of women, single or married, mothers or childless.[23] Particularly in the tobacco factories where the accounts of the grievances suggest sexual harassment (the annoying behavior or *vexations* of an engineer at Bordeaux in 1889, the aspersions cast on the private life of a Dijonnaise by her factory director in 1895, an insult by the Director of Nantes in 1899), one would expect the male tobacco workers to suggest that the women seek the sort of work that they could do in the privacy of their homes: laundry or sewing. On the contrary, there was no suggestion that there was any difference between the attitudes of the male and female tobacco workers: the bosses had committed outrages and should be dismissed or made to apologize.

Why was there such harmony? First, although in other trades women were accused of displacing men by working for lower wages, tobacco and match production had always been majority-female trades. By the 1890s neither had had time to go through technological or other changes which might have influenced sex-based hiring.[24] Most functions were performed indifferently by men or women with the exception of heavy lifting. It was not even clear whether the highest paid jobs were reserved for men, although there were sometimes separate male and female workshops performing the same functions (as there were at the time of my visit in 1983 to the Saintines match factory in the Oise).

Second, no one regarded a job at a tobacco or match factory

as the refuge of a poor girl who should be tolerated rather than cast into poverty. Most of these women had fathers, husbands, or live-in boyfriends whom they eventually married. (See Chapter III: Demography). They chose to work in tobacco or matches; they were conscious of their good wages and the promised pension which no worker, male or female, would sacrifice.

Third, there is evidence that the extraordinary degree of solidarity within the two Federations was further strengthened by family solidarity. We know from the census and personnel records that many match workers had been tobacco workers and that many tobacco and match workers found places for their children in the same factory where they worked. Mannheim claimed that one of the demands of the 1899 Match Workers' Congress was that 70% of future personnel be chosen from the children of the match workers.[25] In my sample of male match workers, approximately one-fifth were living with a girlfriend or wife who worked at one of the Parisian match factories; almost one of every ten had a sibling who did the same. The hiring of an outsider rather than a relative was the cause of a 1901 strike.[26]

Therefore, when we ask why the male members of the Match Workers' were silent on the question of women working outside of the home in the same years when most male trade unionists considered that women's only place was the home, we must take into account that the protection of the home often followed women match workers to work and to union meetings in the person of a father, lover, or brother. When we read of the discussion at the 1908 Congress on possible layoffs due to mechanization, we need not wonder why the men did not follow the lead of men in other trades by suggesting that the women lose their jobs first. Instead, the all-male 1908 Congress voted against any layoffs and then proceeded to the next agenda item: pensions. Obviously, by voting to keep all of the match workers employed, the delegates to the national Congress were also voting for the present and future welfare of their own families.[27]

In fact, the men on the central committee seem to have been even more representative of interconnected match worker families than were the other male match workers. Of the 28 men who served on the central committee between 1893 and 1899 and whose households were found in one or both of the censuses, ten or almost one-third had wives or girlfriends who were match workers. Seven or one-fourth had sisters or brothers with the same experience. Some, such as Jean Dalem, seem to have had both.[28]

We still need to examine why the women tobacco workers were an exception to the rule on union leadership while the women match workers were not. Certainly, in both Federations there was apparent harmony between the women and men. The women were more likely to act as the rank and file organizers and day to day carriers of the union locals while some of the men acted as the public figures. Yet we must explain why women were almost half of the Tobacco Federation's delegates to every national congress in the 1890s; within the Match Federation, they were not.[29]

There were obvious reasons why women were less likely to become frequent participants at union meetings in any trade. Even for those who acknowledged the probability of lifelong participation in the wage labor force and those who had the interest and the patience for meetings, there were material and social constraints. Traveling to meetings and staying out of town cost money; women were paid less than men, so they had less disposable income. Also, we know that some young, single women were under social pressure not to go to evening meetings; dangers to their safety and their reputations also prevented them from traveling out of town alone or in the company of men to a multi-day national union congress.

Consequently, it was not unusual in these years for a majority-female or even an all-female union to be represented by men. This was the case, for example, for the *Dames des cafés-restaurants de Paris* at all of their congresses of the *Fédération de l'Alimentation* between 1904-1912; it was also so in the industrial sector, for example, for the *Ouvrières en peinture-céramique de Limoges*. It was not so for the tobacco workers. From their very first national congress in 1891 until 1913, they had almost as many female delegates as male delegates (See Chart XXIII: Numbers of Male and Female Delegates to Tobacco Federation Congresses 1891-1913). Their presence was not symbolic because they volunteered to take on serious responsibilities.[30] For example, after the Congress of 1891, a delegation of five men and two women presented the demand for official recognition of the Federation to the *Directeur-Général* of State Manufactures. Women were also regularly elected to the central committee. At the 1892 Congress, it was suggested that the presiding officer at each session be a man but the recording secretary be a woman. Female delegates from Marseille and Toulouse objected; therefore, one session was presided over by a woman and some of the recording secretaries were men. The women also took part in all of the discussions, not only on typically female concerns such as daycare facilities and the dues structure, but also on the retirement age and the eight-hour day.

Certainly, within the Tobacco Federation, material and social conditions for participation in union congresses were more favorable to women than they were in other national unions. Also the Tobacco Federation had a higher percentage of women (89% between 1891-1913) than the Match Federation (65-70% between 1893-1900),[31] therefore a larger pool to draw from. Most important, the women tobacco workers were older than their counterparts in the match industry. Thus, the former were less likely to have small children whom they could not or would not leave for several days of a union congress.

CHART XXIII: NUMBERS OF MALE AND FEMALE DELEGATES TO TOBACCO FEDERATION CONGRESSES, 1891-1913

Year of Congress	Female Delegates	Male Delegates
1891	13	13
1892	14	19
1894*	12	18
1910	17	30
1911	16	24
1913	24	25

Figures from Guilbert, 92-99.
* No delegates list found for intervening congresses.

Tempting as it is to hypothesize that the women tobacco delegates were the relatives of male delegates, there is not enough information to confirm or deny this. Unlike the women delegates to socialist congresses, who were often cited in Jean Maitron's biographical dictionary of the French workers' movement as the wives of militants, there is insufficient evidence linking the women tobacco delegates to male delegates or to other men who were active in union or political groups. However, unlike the women who attended socialist congresses and were identified in the minutes as *citoyenne*, women tobacco delegates were usually called *Madame*. As the women delegates to the congresses of other trades were called either *Mademoiselle* or *Madame*, rarely *citoyenne*, this suggests that the recording secretaries were not merely filling a blank with a polite title but that the tobacco delegates were indeed usually married women. Perhaps the Tobacco

Federation's success in recruiting many women delegates lay partly in the choice of married women whose reputations would be less likely to be damaged by out of town trips and long meetings in the company of men. In addition, many (12 to 24) women delegates attended each of the national Tobacco Federation congresses between 1891 and 1913; therefore, the women could travel and be housed together.

However, what about the conflict between their responsibilities to their children and to their union? Let us remember that my inquiry into the alleged reproductive difficulties of the match workers revealed that most of them terminated their child-bearing by the age of 35.[32] Census figures indicated that the women tobacco workers were older than the women match workers. Sixty-four per cent of the tobacco workers were older than 35 while only 29% of the match workers were that old in 1901. Assuming that child-bearing patterns among the two groups were the same, there would have been a pool of approximately 420 women match workers who were free to attend union congresses, if their inclinations and talents so disposed them, compared to more than 10,000 women tobacco workers.

Naturally, if this assumption is correct, the women match workers would have begun to imitate the women tobacco workers in leadership roles in the years after the white phosphorous crisis. By then a larger percentage would have completed their child-bearing. In support of this hypothesis, it is worthwhile to notice that at the 1903 Match Federation Congress, the question of daycare was discussed for the fourth and last time.[33] At the 1911 Congress, for the first time, three of the twenty delegates were women. *Mesdames* Caire and Pantasso had traveled to Paris from Marseille; Marthe Laboudigue (no information available on marital status) had come from Bègles. Their arguments were not especially militant but neither were those of the men.

By 1911 the women who took part were no longer exceptional labor (and socialist) militants like Marie Jay of Marseille or the relatives of labor (and anarchist) militants like Madame Ménard of Trélazé but rank and file women workers who had grown into leadership gradually. This, then, is the reason for the low level of leadership displayed by women match workers during their struggle of the 1890s. Ironically, they elected men as union leaders and allowed them to brandish the argument of reproductive hazards while the women were busy bearing and caring for children.

ENDNOTES CHAPTER VIII

1. In 1900, the first year for which national figures on trade union women are available, women accounted for only 5.3% of union membership, although they were one-third of the working population. Madeleine Guilbert, *Les femmes et l'organisation syndicale avant 1914*, Paris: CNRS, 1966, 29.

2. Often in majority-female industries, the majority of union members were male.

3. The Match Workers' Federation, which had only 1,400 women members scattered over five *départements* was too small to be reflected in these statistics.

4. The government fell over this very question on May 22, 1894. The Merlin-Trairieux bill refused the right to unionize or strike to arsenal and railroad workers. After Minister of Public Works Jonnart was questioned in the Chamber of Deputies, the bill was voted down. One hundred progressive republicans, Jonnart's own party, voted against his view. Goguel, ibid., 73. The *Fédération nationale des Travailleurs réunis des Arsenaux de la Marine de l'Etat* was founded in 1900. The *Fédération nationale du personnel civil des Etablissements militaires de la Guerre* was founded in 1901.

5. Charles Mannheim, ibid., 298.

6. "The decade 1879-1889 was marked by the return of the Communards from exile, by the formation of the first socialist parties and labor unions, by the troubles of Montceau-les-Mines and Lyon (1882), by the strikes of Decazeville and Vierzon (1886). Many signs announced the appearance in France of the social question in its modern form as the labor question....Hypnotized by political and constitutional questions, conservatives and republicans remained blind to the social question and failed to see the signs which already proclaimed its gravity." Francois Goguel, *La politique des partis sous la Troisième République*, Paris: Seuil, 1958, 70.

7. Mannheim, ibid., 418.

8. In the high strike years of 1893-1894, women represented only 11.4% of strikers. Guilbert, ibid., 204.

9. Mannheim, ibid., 416-422.

10. Mannheim, ibid., 298.

11. Mannheim, ibid., 418. Rather, management interviewed each striker individually before readmitting her or him.

12. Mannheim, ibid., 424. However, on page 296 Mannheim wrote that the first tobacco union started in Lyon in 1887 and that the Marseille union was actually formed the following year.

13. In none of the sources was Ducros' first name given.

14. The reasons of the Marseille president were not mentioned. Jay gave her own reason for allegiance to the Federation as a desire to build socialism: "Socialism is the goal of workers and it won't be individualists who prevent the movement." Jean Maitron, ibid., volume 13, 105. Her use of the word individualists (*individualités*) may be significant of a split between an anarchist president and a socialist majority. Again whether the workers were animated by any particular political philosophy and whether the shift of power to Paris was indicative of the victory of one tendency is a question for further research.

15. We know that in the match industry as well the impetus came from Marseille. The first match local formed in Marseille in 1890; Pantin and Aubervilliers were only the fourth and fifth to join, after which the locus of power in that Federation also shifted to Paris.

16. Mannheim, ibid., 428.

17. The leadership of the Tobacco Federation must have been disappointed when the Match Workers' refused to merge their 2,000 members with the 20,000 members of the Tobacco Federation. Efforts to merge at this time and again in 1901 were thwarted by the Match Workers' desire to have more than one page of the projected common newspaper. *L'echo des tabacs* started in 1895, *L'echo des allumettiers* three years later.

18. Pierre-Joseph Proudhon, *La pornocratie ou les femmes dans les temps modernes*, Paris: A. Lacroix, 1875.

19. For a fuller discussion of the problems as they were evoked at trade union congresses, see Guilbert, ibid., 188-193.

20. Guilbert, ibid., 65 citing *Compte rendu analytique* in *L'ouvrier Chapelier*, no. 2 (2 November 1884) - no. 11 (1 March 1885).

21. 1,850 in favor of the admission of women, 5,633 against, 970 abstentions. Guilbert, ibid., 59.

22. This resolution passed narrowly by a vote of 74 in favor, 62 opposed, and 22 abstentions. Guilbert, ibid., 63. Summary of *Compte rendu du dixième Congrès national de la Fédération francaise des Travailleurs du Livre*, Paris: Imprimerie Nouvelle, n.d. However many of the union's locals were not at all resigned

to validating the presence of working women in their trade. In 1912, the Lyon local refused to accept the membership application of Emma Couriau, a typesetter with 17 years of experience, who was already working at union wages. In addition, her husband's name was struck from the membership roles for allowing her to apply. The decision was appealed to the General Association of Lyon typographers who upheld it by an overwhelming vote of 300 to 26 with 11 abstentions. Guilbert, ibid., 63.

23. There is evidence of concessions to those who wanted sex-segregated or sex-mixed unions. At Orléans the *Syndicat des ouvrières aux tabacs de la manufacture d'Orléans*, founded in 1892, was an all-female union local with a membership of 150. There was a second Orléans tobacco union founded in 1894 with a mixed male and female membership of 30. It was not part of the Federation. The *Syndicat des ouvrières des manufactures de tabacs, section d'Orléans*, founded in 1898, with 150 women members was obviously the old union renamed to reflect federation. At Toulouse as well the mixed-sex union founded in 1891 was complemented in 1895 by a larger all-female union. Since the Orléanais and the Toulousains held their meetings at the *Bourse du travail*, it seems that they were like-minded in their aims but some members preferred sexually segregated meetings.

24. Some accounts describe cigar rolling as a skilled male occupation, but all agree that the more mechanized tasks were performed indifferently by men and women. In the 20th century, the percentage of women in French tobacco and match manufacturing has declined because of the preferential hiring of veterans.

25. Mannheim, ibid., 346.

26. Mannheim, idem

27. Guilbert, ibid., 104 citing *Fédération des ouvriers et ouvrières des manufactures d'allumettes de France, Neuvième Congrès du 29 juin au 5 juillet 1908*, Paris: Maison des Fédérations, 1908, 29.

28. We know that his wife, Marie Lontz, was a match worker. There was a Catherine Dalem five years his senior, who may have been his sister. She in turn was married to cental committee member Hippolyte Grégoire. As another example, Louis Barthélemy was nineteen years older than Louise Barthélemy, who worked making match boxes. Perhaps when 18-year old Louise took unauthorized

leave during the match strike of 1895 although her workshop was not on strike, she was simply supporting her relative's cause.

29. The only woman delegate to a national Match Workers' Congress in the 1890s was Madame Ménard from Trélazé. She attended the 1894 Congress in Pantin-Aubervilliers. See Chapter IV: "Four Strike Years" for her possibly unique qualifications.

30. As we know, Marie Jay and Marie Deleuil of Marseille had been among the organizers of the locals at Riom and Orléans, where, Jay claimed, by joining before the men, the women had "once again shown the right way to the men who think themselves the stronger ones". At the 1892 Congress, Deleuil initiated a motion to maintain the level of dues in order to be ready for anything. In the course of the debate with a male delegate from Toulouse, she specified that she meant to keep a warchest (*caisse de résistance*) in case of strike. Guilbert, ibid., 95 citing from *Deuxième Congrès de la Fédération des ouvriers et ouvrières des manufactures de tabacs*, Paris: 1893, 199-201.

31. Guilbert, ibid., 93-99 and 465.

32. Let us recall that 89% of the female match workers in my sample had completed their families by that age according to the 1891 census; likewise, 90% had done so by the 1896 census. These calculations are based on the assumption that the tendency to complete child-bearing by the age of 35 continued until 1901, the first year for which figures are available comparing the ages of women tobacco workers and women match workers. République Francaise, Ministre du Commerce, de l'Industrie, des Postes et Télégraphes, "Résultats statistiques du recensement général de la population effectué le 24 mars 1901", Tome I, 2e. partie, Tableau VI, Paris: Imprimerie Nationale, 1904, 80.

33. Guilbert, ibid., 103.

CHAPTER IX: THE FRENCH STATE AND THE MATCH WORKERS: GENERAL AND CONCRETE CONSIDERATIONS

After October 1898, the Ministers of Finances set about finalizing the necrosis crisis and taming the Federation. On the issue of compensation for necrosis, the Federation had clearly won. However, the Administration had learned two lessons. The first was that procrastination on health and safety could be costly in both money and public reputation. The second lesson was that the Federation members had made too much use of their special status as government workers in the Republic. To remedy the health and safety problem, management therefore proceeded to clean house of all the remaining high risk workers. To tame the Federation, the Administration redefined worker-Federation-management relations in the hopes of avoiding any repetition of the power struggle which they had just lost.

By the end of 1898, the *Direction-Générale des Manufactures d'Etat* was trying to cap any further financial concessions for medical reasons. Spokesmen told Federation representatives that medical benefits were already quite generous: three months of benefits at one franc sixty centimes per day for men and one franc ten centimes per day for women in the *département* of the Seine, slightly less in the provinces. If a worker required another three months of treatment, he or she was eligible for 50% or more of average wages. In the following 12 month period the worker was again eligible for six months of payments.[1] Administration spokesmen kept stressing that their terms were much more generous than those of the workers' compensation law of April 9, 1898: the Ministry of Finances paid benefits from the first day of sickleave, whereas the workers' compensation law provided for payments only from the fifth day. Nonplussed, the Federation kept asking for, and in some cases receiving, sick pay equal to 100% of wages plus free medicine.

Indeed, when the union brought up the case of an individual who was not cured within six months, the Administration spokesmen tried to make an exception for that individual without giving the Federation a precedent-setting victory. For example, in January 1899, Federation officials argued on behalf of two women who were suffering from especially severe cases. One Parisian had been hospitalized for 22 months for a gastric ulcer, which her doctor had diagnosed as phosphorous-related. Another woman at Trélazé had been totally disabled since September 1897 by the secondary effects of an advanced case of necrosis. Although the Administration agreed to supply them each with a liter of milk per day, the Trélazean died in April 1899.[2] On the subject of compensation, a letter dated January 1899 from the Minister of Finances to the Federation is revealing in its language. He promised *à titre tout à fait exceptionnel*, as a totally exceptional measure, that the highest possible compensation would be provided for any remaining dental problems at Pantin-Aubervilliers; then he hedged by adding - if his budget so permitted.[3] He was twisting and turning in an effort to simultaneously promise that the Ministry would do the maximum for all of the victims but refrain from setting precedent.

Next, in order to clean house by pensioning off all the remaining high risk cases, the Ministry of Finances secured a special allocation of 200,000 francs in June 1898. These funds permitted the administration of the match factories to count the years which pension applicants had worked under the *Compagnie Générale*; those who could show 25 years of service were eligible immediately.[4] Those workers who had fewer than 25 years of service but were diagnosed as incurable were also eligible for immediate pensions of at least two-thirds of the previous year's wages. Finally, those who were not ill but were considered high risks would receive a minimum of 600 francs for the men or 400 francs for the women.

Skimpy as these pensions may seem, they proved very costly to the Administration. For example, one worker named Emile Loegel developed a slight case of necrosis in 1894, after four years at Pantin-Aubervilliers, was carried on the payroll for 27 months, and was then awarded a pension of 950 francs per year. He collected from the time of his withdrawal from the factory at the age of 32 until his death at the age of 64 in 1929. After his death, his widow applied for a necrosis widow's pension. This last request was denied on the grounds that his death could not have been phosphorous-related. This extreme case illustrates how expensive and long-lasting the Administration's liability could have become had it not banned the poison in 1898. It also illustrates the urgency for the Administration of setting definite figures on the settlements of those workers who could legitimately claim necrosis-related disability pensions.

By May 1899, with several workers still receiving sick pay in the hopes of either recovering or qualifying for a pension, the Director-General of State Manufactures issued this circular:

"It is intolerable for a person who has not performed any work

165

for years to continue to be carried indefinitely on the lists of workers in need of (financial) help."[5]

From that month, any match worker who had not worked within the previous 24 months had either to provide proof that he or she was suffering from necrosis or return to work immediately. There would be no more sick pay.

Were some workers indeed exaggerating every minor symptom in the hopes of profiting from pensions? *Le Petit Bleu*, a newspaper which was sympathetic to workers, gave that impression in this article of October 1899:

> "A pretty twenty year old match worker came to us and joyously said: 'Oh! If you only knew how happy I am!..Do you know Big Jeanne?...a match worker for six years, she has finally got her 1,500 francs (pension) for her phosphorous disease, necrosis I think they call it ...Luck, huh? We all get it sooner or later...Well me, I've been waiting for it for a year. Every morning I've been checking my gums, my teeth, but nuts, we've got great teeth in my family...Well, this morning while I was rummaging around in my mouth, what did I find but the famous black dot that's the beginning of it all. Did I jump for joy! You know, in a year, maybe eighteen months...I'll have my 1,500 francs so I can marry Ernest who's been waiting for this..."[6]

The factory dentist also complained that workers were irritating their gums with pins in efforts to develop the symptoms.[7] In addition, some private doctors were still too sympathetic to suit the State Manufactures. Dr. Courtois-Suffit, who had been brought from the Marseille factory to Pantin-Aubervilliers at the end of 1896, reported:

> "Because necrosis had received so much publicity, especially in the political newspapers, it assumed for both the workers and the doctors the qualities of a terrifying ever-present specter... This...continued to haunt the minds of many doctors and was certainly the greatest difficulty I had to overcome at the beginning of my service at the Pantin(-Aubervilliers) factories. A worker who made an appointment with any dentist whatsoever and showed any lesion of the mouth whatsoever had only to say that she was a match worker to get a certificate of necrosis..."[8]

However, as we know, in May of 1899, the Director-General issued a circular to his factory directors declaring that the danger of necrosis was a thing of the past. In June he further specified that the new workers' compensation law of April 9, 1898 was not applicable to

"a slowly evolving malady whose cause might be attributable to occupational disease..."[9]

In other words, if a worker was suffering from toothaches or the vague symptoms of phosphorism, there would be no compensation or pension. If indeed a worker had a case of necrosis, there would be compensation but certification at the most troublesome factories, Pantin and Aubervilliers, would come only from Dr. Courtois-Suffit.

Despite the workers' ongoing concern over latent necrosis cases, they emerged generally satisfied. Once the medical emergency had passed and the last financial settlements had been made, however, the Administration started to apply some lessons learned during the struggle in order to curtail the power of the Federation.

After 1898, there were changes in the hiring policy to decrease the workers' security and new rules to limit the Federation's scope of action. The mass layoffs which had been discussed in 1896 were threatened again in 1900. By that date the State engineers Sévène and Cahen had succeeded in producing enclosed machinery similar to the models of the Diamond Match Company. Although the massive layoffs that the workers feared did not take place, there was increased use of temporary workers who could be laid off at will and accumulated no pension rights.[10] Disturbing as this was to the strength and unity of the Federation, a certain amount of automation could only have been expected because of its successful use in other countries; the necrosis crisis had simply provoked the Administrators of the French monopoly to invest in new machinery sooner rather than later.

More direct efforts to curtail the Federation's power appeared in October 1900 in an official circular from the Minister of Finances Joseph Caillaux to the *Direction Générale des Manufactures d'Etat*. This circular offered the first definition of the hierarchy to be observed in the grievance procedure. With reference to strict respect for the law of March 21, 1884, which had legalized labor unions, Caillaux made clear that he would not be bothered or have the *Direction Générale* bothered by the Federation for minor grievances.

"The experience of these last years has brought forth the need for specific rules on union-management relations within the State Manufactures in order to avoid any repetition of conflicts, no doubt involuntary, but nevertheless harmful to the good management of the monopoly and to workshop discipline..."[11]

Henceforth, the *Direction Générale* was to meet only with representatives of the Federation, not with representatives of the Pantin-Aubervilliers locals. Since these two were identical, the Minister of Finances found it necessary to add that such meetings were to be limited to questions of general interest concerning all seven factories. Thus, there would be no more appeals made to the top for the pensions of individual workers. Second, all demands were to be in writing, a rule designed to hamper many union members who had come of age before the compulsory elementary school law of 1884.[12] Thus, there would be no more storming a factory director's office for an immediate confrontation. Even when the Federation leadership abided by these rules, the *Direction Générale* was not to embarass the Directors of individual factories by answering immediately in person, but rather was to respond through an intermediary. In this way there would be no more awkward three-way negotiations such as Ernest Deroy had forced in 1893 with the Director of Pantin-Aubervilliers, Descombes, and the Director-General of State Manufactures, Pradines.

On October 30, 1900, the *Directeur-Général* in turn sent out his circular to the individual factory directors. From that date forward, there would be no union business conducted on the work premises: no posting of bills, and no soliciting or receiving of grievances without prior authorization. A demand for a grievance hearing was to be submitted in writing to the factory Director who might authorize a visit to the workshop in question by a single union representative in the company of a supervisor.

"It is understood that demands on the Administration which are not sent through channels will be disregarded."[13]

Here, finally, was an effort to stymie the sort of indefatigable Federation militance which had won the match workers their victory in 1898. He reminded the Federation delegates that they were to be polite, deferential, and respectful at all times. Any breach of courtesy would be reported. Obviously, the Administration intended to prevent any more meetings in the style of Ernest Deroy and Jacques Aschbacher who had been known for their disrespectful and violent language.

Certainly these regulations confined the Federation but not as much as the Minister had hoped. The individual match workers did not lose their habit of turning to the Federation for any and all grievances; as late as 1925 an observor wrote that the union representatives still sometimes moved freely from workshop to workshop and called impromptu meetings without asking permission. The match workers continued to use their dual rights as citizens of the Republic and State workers to protest to their elected officials. They also continued to threaten strikes. Pantin-Aubervilliers walked out briefly in 1898, 1900, 1901 and 1902; Aix and Marseille struck in solidarity in 1901. The Seine police report on the 1903 Congress mentions a protest to Parliament on layoffs and unresolved phosphorous-related pensions.[14] And they continued to be known for their political volatility in support of other working class causes as this 1909 police report confirms:

"The match workers are considered the most agitated (*remuants*) State workers..."[15]

The white phosphorous struggle was recuperated by a government eager to control the working class in general and its workers in particular. The Match Workers' Federation had extended its influence over workplace health and safety but had paid a price in diminished freedom of action. Still, the Ministry of Finances did not succeed in taming the Federation, because its militant tradition had become set in the course of the struggle.

It is surprising that the match workers won under these conservative ministries of moderate Republicans and Catholic conservatives. These were not years of social reform. The men in power saw organized labor as a threat to stability, therefore as a proper area of society for surveillance by the Ministry of the Interior. Also, these were years of anarchist bombings and efforts to detect anarchism in every workers' meeting. This explains the plentiful police archives on the match workers but the near absence of papers concerning them in the Archives of the Ministry of Finances. Furthermore, the Ministers of Finances held office for a short time. There were eight Ministers of Finances between 1890-1899. It can be argued that even had they wished to improve the lot of labor, they could have hardly learned the workings of the match factories much less improved them before their ministry fell from office.

Why then did the match workers win? Part of the answer is that the Ministry of Finances underestimated how strong a position the match workers were in. Finances had more serious problems in these years than one badly understood disease in one small industry. Ministers of Finances were inexperienced in labor-management relations: they were chosen for their experience in finance. Not only were they in office briefly, but they concentrated their efforts on major budgetary problems, leaving the day-to-day operations to the *Direction Générale des Manufactures d'Etat*, who also underestimated the situation.

Pierre Tirard was the Minister during the 1893 strike. As Minister of Finances from 1882-1885 and 1887-1888, he had not been sympathetic to direct administration of the match factories; the strike only confirmed his opinion. Although he conceded benefits which were to become unexpectedly costly in later years, at the time they seemed minor and reasonable.

Paul Peytral, who took office immediately after the strike, had also favored returning the match industry to the private sector during his previous term at Finances in 1889.

Next was the colorless Auguste Burdeau. Although he was very knowledgeable on financial matters, he had no experience with organized labor. The question of union rights for State workers was before the Chamber during his five-month ministry, but he apparently showed no leadership on the question. Representatives from the Match Workers' Federation were meeting with delegates from the railroad workers' union to protect their rights when the government fell over this very question.[16]

Alexandre Ribot was the Minister of Finances during the long strike of 1895. He had at the same time the responsibility of the Premiership. His conservative politics may explain his refusal to be bothered with direct negotiation or a special allocation for wages during the 1895 strike. He did, however, visit the Pantin-Aubervilliers factories in September 1895, where he promised a variety of expensive improvements. At that time, it was to his advantage to draw attention away from government complicity in the repression of the Carmaux glass strike and the maladministration of the *Chemins du Fer du Sud*. This last scandal brought his government down on October 28, 1895.

Next, under the five-month all-Radical ministry, Paul Doumer was busy pressing a controversial budget which would have eliminated the public debt, introduced a graduated income tax, and provided public assistance to the elderly. The Chamber rejected it. In the meantime, although Doumer expressed verbal sympathy for organized labor, he could not have secured the support which would have been necessary to make expensive changes in the match industry. Instead he looked into rationalizing the industry by automation. Fortunately for the match workers, automation was even more expensive.

Finally, the Minister of Finances who had the longest term of office, 26 months, was Georges Cochery. He served under Premier Méline, an enemy of labor. He was chosen because of his long experience with State finances and his resolute opposition to an income tax. This opposition gave him all the more reason to try to make State industries smooth-running and profitable. Finances had never realized that white phosphorous was much more an organizing tool than a health problem. Therefore, Cochery, like his predecessor, threw the problem at the medical profession. However, unlike his predecessors, he was faced with a production crisis when one-third of Pantin-Aubervilliers went on sickleave. This permitted him to solve the problem expensively but once and for all. Hence, there was the precipitous change of company doctors and the examination of all the Pantin-Aubervilliers workers. Second, Cochery's influence may be seen in the encouragement which was finally given to inventors and the fact that the two engineers on the payroll of the Ministry of Finances were given the research time and the laboratory materials with which they found the altenative.

Ironically, it was these ministries, which did not distinguish themselves for social intervention, which were to receive the credit of historians for this important precedent in occupational health. Rather, the Ministers of Finances, by underestimating the situation, left just the openings which these unusual workers needed. By failing to realize early that necrosis was a labor organizing issue as well as a legitimate health problem, Finances allowed the Match Workers' Federation to grow in experience and self-confidence. By dumping the problem onto the medical profession but refusing to abide by their recommended ban, the Ministry presented an opportunity for an enterprising doctor to step into the role of advisor to a labor union. Particularly important was that all of the Ministers of Finances failed to gauge the public mood concerning working women of child-bearing years. In this way, the Ministry allowed the match workers to shame it at a time when there were no studies to prove that the women match workers' reproductive health was not endangered.

But weren't these the ministries which took pioneering steps in workplace legislation? They passed the law of 1892 on hours, the law of 1893 on sanitary conditions of workplaces, and the 1898 law on workers' compensation. The idea of social responsibility for the ravages which industrial capitalism had brought to all men, women, and children of the working class was current in the minds of all elected officials, economists, and large manufacturers. Although many continued to blame the workers for their plight, they did agree sufficiently to pass limited and cautious legislation. They did so, as I argued earlier in the text, without adequate enforcement in the cases of the hours and sanitary conditions law. The workers' compensation law benefited insurance companies and protected business more than it provided compensation genuinely equivalent to the losses of limbs and lives. Only in the case where the State was the employer could its representatives not publicly regret the ravages of industry but refuse to interfere too heartily with private property. The match factories were public property and therefore a showcase of labor relations, although the Ministers of Finances talked as if they were the private property of the Ministry and merely a source of income. Therefore, public pressure was brought to bear, especially at the time of the 1896 sickleave crisis. Again the match workers shamed the Administration.

It is not for these reasons that the match workers won. It is for these reasons that their victory was possible. What turned possibility into reality were the actions of the match workers themselves. In a time of ministerial instability, nascent respect for the opinions of medical science, republican consolidation, and national concern over the birthrate, this group of workers made several unusual accomplishments. They organized labor unions of industrial workers in the government sector. They managed harmonious trade union relations between male and female members. They succeeded in lobbying and gaining the attention of the press. They established *de facto* workers' compensation for an industrial illness before there was *de jure* workers' compensation for most industrial accidents. Finally, they secured the ban and replacement of a toxic chemical, which ban was then adopted by the other match manufacturing countries. In so doing, they simultaneously secured the removal of a cheap, readily available poison from the household of every match consumer. For these last two accomplishments, the match workers and match consumers of Europe and North America owe them gratitude.

Their example was not imitated by many, which is why labor historians until recently have considered the organization of women in industry to be an infertile ground for research. On the contrary, this study shows that advanced demands could be won and viable mixed unions could be maintained by entirely unskilled, blue collar women of the working class. They were not the articulate members of a class in transition. They were not the genteel daughters of New England farmers who formed the Female Labor Reform Associations in the Massachusetts spinning mills of the 1840s.[17] They were not the educated daughters of the French urban petit bourgeoisie, like the leaders of the all-female P.T.T. union, the *Union des Dames de la Poste* in the first decade of this century.[18] In organizing and sustaining the Match Workers' Federation, they established a model by which historians can judge the efforts of other women in mass production and by which labor organizers can recognize some of the conditions which favor growth. For example, the victory over white phosphorous failed to occur first in England because the English women match workers belonged to a newborn union which apparently did not develop viable leadership. It was favored with public support only as long as Annie Besant attached her personal popularity to the cause.[19] Likewise, the victory failed to occur first in the U.S. because the Diamond Match Company, which held the monopoly, had the power to flatly deny the existence of any cases of necrosis.[20] Labor historians too owe gratitude to the French match workers.

> "Whether it is by direct means (strike, petition, demonstration) or indirect means (by pressure on Parliament or on the government) organized workers took a large role in bettering their own general living and working conditions, a necessary precondition so they could benefit from scientific advances at the time of their discovery."[21]

As Rolande Trempé wrote in the above quotation, many workers whose names are not known to history were instrumental in struggles whose stories have not been told. History is written about those who turn their desires into actions, but historians are usually able to find only what has been written by those who turned the desires of the inarticulate into words. Thus, historians have neglected the majority of the actors in events of national and international importance. For example, we know little of the working class individuals, much less of the women, who defended Paris against the *fédérés* in 1871 despite the efforts of Edith Thomas to find their names and give them characters.[22] We know little of those who rose up in 1848 or 1793 despite the efforts of Georges Rudé and Albert Soboul.[23] Even more have historians neglected the rank and file workers in the modern industrial setting. This thesis is a contribution to the growing body of social history which attempts to fill that gap.

ENDNOTES CHAPTER IX

1. This was an improvement on the terms of 1895, before the medical crisis. In the 1895 budget, there were similar levels of sickpay only for permanent workers (*inscrits à titre définitif sur registres matricules*). Payments started on the fourth day of absence. After three months the compensation was reduced by half.

2. *L'Echo des allumettiers*, March and April, 1899. The dead woman was Jeanne Colognard.

3. ibid., March 1899.

4. In 1892, the retirement age was 65 with 30 years of service; in 1897, the Administration decided to count the years worked for the *Compagnie Générale* as far back as 1875. On April 17, 1901, the minimum number of years of service for disability retirement was reduced to 20 years. Direction Générale des Manufactures d'Etat, *Circulaires 1882-1901*, Archives of the Saintines Match Factory, Circulars #45, September 13, 1897; #49, June 14, 1898; #67, April 17, 1901 (hereafter called Direction Générale, *Circulaires*.)

5. Direction Générale, *Circulaires*, #52, May 15, 1899.

6. "Industries homicides", *Le Petit Bleu*, October 18, 1899.

7. Dr. George Cunningham, "Report to the Secretary of State for the Home Department on the Use of Phosphorous in the Manufacture of Lucifer Matches", Great Britain, *Parliamentary Papers*, London: Eyre and Spottiswoode, 1899, 685.

8. Dr. Courtois-Suffit, "Le phosphorisme professionnel", *La Presse Médicale*, May 3, 1899, 205.

9. Direction Générale, *Circulaires* #52, May 15, 1899; #53, June 29, 1899.

10. In the archives of Pantin-Aubervilliers, which were stored at the Saintines match factory, I found a list of temporary workers who were called on for periods of a few weeks or months between 1897 and 1913, until they either found permanent places (*titulaires*) or were laid off due to the elimination of a job (*suppression d'emploi*). One example is the work record of Sophie Guyaux: she became a match worker at the age of 16 in August, 1897, left after two months, returned ten days later to work for four months, reappeared five years later to work for one month in 1903, four months in 1904, and five months in 1905. She then became a permanent worker. According to one source, the policy of using more auxiliaries and fewer permanent workers

was instituted in 1900. *La CGT et le Mouvement syndical*, Paris: L'Action Internationale, 1925, 321; Direction Générale des Manufactures d'Etat, *Registre Matricule pour l'Inscription des Ouvriers et Ouvrières Temporaires*, 48, Manufacture de Pantin-Aubervilliers, March 1890; Direction Générale des Manufactures d'Etat, *Circulaires*, ibid., #67.

11. Minister of Finances Joseph Caillaux to Direction Générale des Manufactures d'Etat in Direction Générale, *Circulaires*, ibid., #62, October 7, 1900.

12. This rule must have impeded the expression of many grievances because we know that in 1895, 16-year old Léon Jouhaux had been one of the few workers who was able to write tracts and take minutes at meetings. Jean Maitron, ibid., Volume 13, 123.

13. Directeur-Général Emile Jacquin, "Ordre de Service Concernant les Rapports du Personnel Ouvrier avec ses Chefs", Direction-Générale des Manufactures d'Etat, *Circulaires*, ibid., #63, October 30, 1900. In December 1899, two union militants from Pantin-Aubervilliers had received punitive layoffs for making the rounds of workshops without authorization. Several turbulent union meetings were held; the Federation asked Premier Waldeck-Rousseau to arbitrate; he referred the matter back to Finances. Circular #63 was probably the Directeur-Général's effort to have the last word. APP Seine, 1.408 bis.

14. Two thousand match workers struck for 17 days when a woman union member lost her job to the protegée of a supervisor. The union member was reinstated. *La CGT et le Mouvement syndical*, ibid., 317, 320.

15. Prefect of Police, April 1, 1909, "Allumettiers 1899-1929", AN F7 13635; *La CGT et le Mouvement syndical*, ibid., 322.

16. Charles Jonnart, 1857-1927, was the Minister of Public Works who refused the right to organize to those on government payrolls.

17. Philip S. Foner, *Women and the American Labor Movement*, London: The Free Press, 1982, 18-27.

18. Susan D. Bachrach, *Dames Employées: The Feminization of Postal Work*, Haworth, 1984.

19. Lowell J. Satre, ibid., 26.

20. Only after the French match workers had provoked the practical resolution of the problem could reforming intellectuals including Dr. John R. Commons of the University of Wisconsin Economics Department and Dr. Alice Hamilton, an expert on occupational

health, catch public attention. In 1906 they founded the American Association for Labor Legislation, which soon published John B. Andrews' study of 150 necrosis cases. In 1910, his report was published by the U.S. Bureau of Labor. In 1911, the U.S. adopted phosphorous sesquisulfide matches and, in the next year, passed a white phosphorous match act, the Esch-Hughes Act. Barbara Sicherman, editor, *Alice Hamilton, a Life in Letters*, Cambridge: Harvard, 1984, 153-160; *Encyclopedia of the Social Sciences*, New York: Macmillan, 1967, Volumes 9-10, 204.

21. Rolande Trempé, "Lutte des travailleurs et améliorations de la santé", in A.-E. Imhof, *Le Vieillissement*, Lyon: Presses Universitaires, 1979, 141.
22. Edith Thomas, *The Woman Incendiaries*, New York: Braziller, 1966.
23. Georges Rudé, *The Crowd in History*, New York: Wiley, 1964; Albert Soboul, *Les Sans-Culottes Parisiens en l'An II*, Paris: Clavreuil, 1958.

BIBLIOGRAPHY

DOCUMENTS

Archives Nationales

F^7 13602	*Bourses du travail, Finistère à Gironde.*
F^7 13611	*Bourses du travail, Oise à Pyrénées-Orientales 1893-1919.*
F^7 13635	*Allumettiers 1899-1929.*
F^{12} 4951	*Allumettes chimiques, fabrication, 1864-1889.*
F^{22} 333	*Durée du travail enfants et femmes, Reglementation du travail: legislation 1847-1936.*
F^{22} 528	Bureau de l'Association internationale pour la Protection Légale des Travailleurs, "Mémoire explicatif sur l'interdiction de l'emploi du phosphore blanc dans l'industrie des allumettes."

Archives Départementales

AD Maine-et-Loire	71 M 2-5, *Grèves.*
AD Oise	#105 Saintines, "Procès-Verbal de la Réunion du 12 avril, 1896."
AD Paris D2M8	*Dénombrements 1891 and 1896, Pantin and Aubervilliers.*

Archives de la Préfecture de Police

B <u>A</u>/400	*Enquête parlementaire des Classes Ouvrières 1872.*
1.408 bis	*Allumettiers.*

Published Documents

Bottin. *Annuaire-Almanach du Commerce et de l'Industrie,* Paris: Firmin-Didot, 1865.

Bulletin officiel de la Bourse du Travail de Paris, Volumes I-III, 1887-1889, Paris: Allemane, 1889-1892.

Bulletin officiel de la Bourse du Travail de Paris, Volumes V-VI, 1892-1893, Paris: Allemane, 1894.

Congrès national des ouvriers et ouvrières des manufactures d'allumettes, 1892, 1894, 1896, 1899, 1903.

Rapport général sur les travaux du Conseil d'Hygiène Publique et de Salubrité du département de la Seine, 1840-1896, Paris: Imprimerie Chaix.

Published Documents: France

Direction Générale des Manufactures de l'Etat, *Compte en matières et en deniers de l'exploitation du monopole des allumettes chimiques*, Paris: Imprimerie Nationale, 1895, 1896, 1897, 1898.

Direction Générale des Manufactures de l'Etat, *Circulaires 1882-1901*.

Direction Générale des Manufactures de l'Etat, *Lois, Ordonnances, décrets, décisions et arrêtés ministeriels concernant le monopole des tabacs et le monopole des allumettes chimiques depuis l'établissement du régime exclusif jusqu'au 31 décembre 1899*, Volume 11, Paris: Imprimerie Nationale, 1901.

Direction Générale des Manufactures de l'Etat, *Registre matricule pour l'inscription des ouvriers et ouvrières temporaires à dater du premier janvier 1909*, Number 48, Manufacture de Pantin-Aubervilliers.

Journal Officiel, Chambre des Députés. Débats, November 19, 1889, 88; November 23, 1889, 132; November 24, 1889, 331.

Journal Officiel, Lois et décrets, December, 1908, 8971.

Marseille, Conseil municipal, *Délibérations, 1878, 1879, 1881*, Marseille: Imprimerie Moullot, 1879-1883.

Ministère du Commerce, *Annuaire statistique de la France*, Paris: Imprimerie Nationale, 1893-1914.

Ministère du Commerce, Office du Travail, *Les associations professionnelles ouvrières*, Volumes I-IV, Paris: Imprimerie Nationale, 1899-1904.

Ministère du Commerce, Office du Travail, *Bulletin de l'Office du Travail*, 1894-1898.

Ministère du Commerce, Office du Travail, *Poisons industriels*, Paris: Berger-Levrault, 1901.

Ministère du Commerce, Office du Travail, *Statistique des grèves et des recours à la conciliation survenus pendant l'année*, Paris: Imprimerie Nationale, 1890-1895.

Ministère des Finances, *Cahier des charges pour l'exploitation du monopole des allumettes chimiques (le 7 juillet 1884)*, Paris: Imprimerie de Chaix, 1884.

Published Documents: Great Britain

Parliamentary Sessional Papers, *Use of phosphorous in the manufacture of lucifer matches. Reports to the Secretary of State for the home department by Professor T.E. Thorpe, Professor Thomas Oliver, and Dr. George Cunningham*, London: Eyre and Spottiswoode, 1899.

UNPUBLISHED ARTICLES, THESES, DISSERTATIONS

Bachrach, Susan, *The Feminization of the French Postal Service 1750-1914*, Ph.D. dissertation, University of Wisconsin, 1981.

Cottereau, Alain, "Santé et culture ouvrière. Approche historique: l'exemple de la tuberculose", Unpublished paper prsented to the June 1980 seminar Travail-Santé, U 88, Institut National de la Santé et de la Recherche Scientifique (INSERM).

Desjours, C. "Organisation du travail et santé mentale", Unpublished paper presented to the June 1980 seminar Travail-Santé, U 88, INSERM.

Hildreth, Martha L. "The Foundations of the Modern Medical System in France: Physician, Public Health Advocates, and the Medical Legislation of 1892 and 1893", Presented at the University of California-Riverside, 1980.

Stone, Judith F. "The Discovery of the Working Class: The Origins of French Social Reform Legislation", paper presented to American Historical Association Convention, 1981.

Stone, Judith F. *Social Reform in France: the Development of its Ideology and Implementation, 1890-1914.* Ph.D. dissertation, State University of New York at Stony Brook: 1979.

NEWSPAPERS

L'Echo des allumettiers, 1897-1899.
L'Echo de Paris, 1893-1897.
L'Echo des tabacs, 1895-1901.
L'Eclair, 1895-1897.
Le Figaro, 1896.
La Fronde, 1897-1898.
La Gironde, 1893.
L'Humanité, 1913.
L'Intransigeant, 1893-1897.
Le Journal du Peuple, 1899.
La Justice, 1893-1898.
La Lanterne, 1895-1898.

Le Libertaire, 1900.

La Libre Parole, 1895-1897.

La Marseillaise, 1893.

Le Matin, 1896.

L'Ouvrier des Deux Mondes, 1897-1898.

La Paix, 1896-1897.

La Patrie, 1895-1896.

Le Père Peinard, 1890, 1892, 1893, 1897.

Le Petit Bleu, 1899.

Le Petit Journal, 1896-1897.

Le Petit Parisien, 1893, 1896, 1897, 1904.

La Petite République, 1895-1899.

Le Prolétaire, 1897.

La Question Sociale, 1895.

Le Radical, 1895-1897.

Le Rappel, 1893-1896.

La Revue socialiste, 1885-1896.

Le Sémaphore de Marseille, 1888-1889.

Le Siècle, 1895.

Le Socialiste, 1892, 1893, 1895.

Le Temps, 1886-1897.

Les Temps Nouveaux, 1895-1898.

BOOKS

Adam, Jacques, *Des maladies professionnelles occasionnées par le travail*, Paris, 1908.

Albert, Pierre, *Histoire de la presse politique nationale au début de la Troisième République 1871-1879*, Volumes I and II, Paris: PUF, 1980.

Alfassa, Georges, *Le travail de nuit des femmes et l'interdiction de l'emploi de la céruse et du phosphore blanc dans l'industrie*, Paris: Larose, n.d.

Antoine, Michel; Barral, Pierre; et al, *Origines et Histoire des Cabinets des Ministres en France*, Genève: Droz, 1975.

Ariès, Philippe, *Centuries of Childhood*. New York: Knopf, 1962.

Armengaud, André, *Les Francais et Malthus*, Paris: PUF, 1975.

Arnaud, Dr. Francois, *Etudes sur le phosphore et le phosphorisme*, Paris: Baillière, 1897.

Association Internationale pour la Protection Légale des Travailleurs, *Les industries insalubres. Rapports sur leurs dangers et les moyens de les prévenir, particulièrement dans l'industrie des allumettes*, Iena: Fischer, 1903.

Auzier, Claire and Houel, Annik, *La grève des ovalistes*, Paris: Payot, 1982.

Bachrach, Susan, *Dames Employées: The Feminization of Postal Work*, New York: Haworth, 1984.

Bance, Pierre, *Le syndicalisme ouvrier francais dans la genèse du droit du travail, 1876-1902*, Paris: Thèse en droit, 1976.

Bellanger, Claude; Godechot, Jacques; Guiral, Pierre; Terrou, Fernand, *Histoire générale de la presse francaise 1871-1940*, Volume III, Paris: PUF, 1972.

Bertillon, Jacques, *De la dépopulation de la France et des remèdes à y apporter*, Paris: Alcan, 1911.

——————*De la fréquence des principales causes de décès à Paris pendant la seconde moitié du 19e. siècle*, Paris: Imprimerie municipale, 1906.

Bonneff, Léon and Maurice, *Les métiers qui tuent, enquête auprès des syndicats ouvriers sur les maladies professionnelles*, Volume I, Paris: Bibliothèque d'études ouvrières, 1906.

Bothereau, Robert, *Histoire du syndicalisme francais*, Paris: PUF, 1945.

Bouvier, Jean, *Histoire économique et histoire sociale, Recherches sur le capitalisme contemporain*, Genève: Droz, 1968.

Bron, Jean, *Histoire du mouvement ouvrier francais, La contestation du capitalisme par les travailleurs organisées, 1884-1950*, Volume II, Paris: Editions ouvrières, 1970.

Chastenet, Jacques, *Histoire de la Troisième République, La République Triomphante 1893-1906*, Paris: Hachette, 1955.

Clèdes, Anne-Marie, *Histoire de l'Allumette*, Paris: SEITA, n.d.

Coleman, William, *Death is a Social Disease*, Madison: University of Wisconsin Press, 1982.

La CGT et le mouvement syndical, Paris: l'Action internationale, 1925.

Courtois-Suffit, Dr. Maurice, *La Prophylarie du phosphorisme professionnel*, 1899.

Derme, M. *Le Monopole des Allumettes*, Paris: Rousseau, 1911.

Desoille, H.; Scherrer, J.; Truhaut, R. *Précis de médecine du travail*, Paris: Masson et Compagnie, 1975.

Dolléans, Edouard, *Histoire du mouvement ouvrier*, Volume II, Paris: Colin, 1939.

Encyclopédie d'hygiène et de médecine publique, Volume VI, 1896.

Encyclopédie médico-chirurgicale, Volume II, Paris: n.d.

Encyclopedia of Social Sciences, 1967, s.v. "white phosphorous".

de Félice, Raoul, *Les Naissances en France*, Paris: Hachette, 1910.

Foner, Philip S. *Women and the American Labor Movement*, London: The Free Press, 1982.

Goguel, Francois, *La politique des partis sous la Troisième République*, Paris: Seuil, 1958.

Gras, L.-J., *Historique de l'Armurerie Stéphanoise*, St.-Etienne: n.p., 1905.

Guilbert, Madeleine, *Travail et condition féminine: bibliographie commentée*, Paris: Editeurs de la Courtille, 1977.

——————————*Les femmes et l'organisation syndicale avant 1914*, Paris: CNRS, 1966.

Hamon, A. *La France sociale et politique: 1891*, Paris: Savine, 1892.

Hatzfeld, Henri, *Du pauperisme à la sécurité sociale. Essai sur les origines de la sécurité sociale en France 1850-1940*, Paris: Colin, 1971.

Jolly, Jean, *Dictionnaire des parlementaires francais; notices biographiques sur les ministres, sénateurs, et députés*, Volumes I- VI, Paris: PUF, 1960.

Kinnersley, Patrick, *The Hazards of Work: How to Fight Them*, Pluto Press, 1973.

Knibiehler, Yvonne and Fouquet, Catherine, *L'histoire des mères*, Paris: Montalba, 1980.

Kuisel, Richard, *Capitalism and the State in Modern France*, Cambridge: University Press, 1981.

Laot, Jeannette, *Stratégies pour les femmes*, Paris: Stock, 1977.

Lequin, Yves, *Les Ouvriers de la région lyonnaise 1848-1914*, Volumes I and II, Lyon: Presses universitaires de Lyon, 1977.

Leroy, Maxime, *La Coutume ouvrière: syndicats, bourses du travail, fédérations professionnelles, coopératives. Doctrines et institutions*, Volumes I and II, Paris: Giard et Brière, 1913.

Leroy-Beaulieu, Paul, *L'Etat moderne et ses fonctions*, Paris: Guillaumin, 1890.

——————————*Le Travail des femmes au dix-neuvième siècle*, Paris: Charpentier, 1873.

Levasseur, Emile, *La population de la France*, Volumes I-III, Paris: n.p., 1889-1892.

──────────*Questions ouvrières et industrielles en France sous la Troisième République*, Paris: Rousseau, 1907.

Louis, Paul, *Histoire du mouvement syndical en France, 1789-1918*, Volume I, Paris: Valois, 1947.

McLaren, Angus, *Sexuality and Social Order. The Debate over the Fertility of Women and Workers in France, 1770-1920*, New York: Holmes and Meier, 1983.

Magitot, Dr. Emile, *De la nécrose phosphorée. Rapport lu à la Société de chirurgie de Paris*, Paris: Imprimerie Chamerot, 1874.

Maitron, Jean, *Dictionnaire biographique du mouvement ouvrier francais*, Volumes 10-15, Paris: Editions ouvrières, 1973.

──────────*Histoire du mouvement anarchiste, 1880-1914*, Volumes I and II, Paris: Maspéro, 1975.

Malot, Hector, *En famille*, Paris: Editions G.P., 1960.

Mannheim, Charles, *De la condition des ouvriers dans les manufactures de l'Etat (Tabac-Allumettes)*, Paris: 1902.

Martial, René, *L'ouvrier, son hygiène, son atelier, son habitation*, Paris: O. Doin et fils, 1909.

Marx, Karl, *Capital*, Volume I, New York: International Publishers, 1974.

Mayeur, J.M., *Les débuts de la Troisième République, 1871-1898*, Paris: Seuil, 1973.

Milhaud, Caroline, *L'ouvrière en France. Sa condition présente. Les réformes nécessaires*, Paris: Alcan, 1907.

Montreuil, Jean, *Histoire du mouvement ouvrier en France*, Paris: Aubier, 1946.

Nourrisson, Paul, *Le risque professionnel et les accidents du travail*, Paris: Larose et Forcel, 1891.

Pelloutier, Fernand and Maurice, *La Vie ouvrière en France*, Paris: Schleicher, 1900.

Perrot, Michelle, *Les Ouvriers en grève*, Volume I, Paris: Mouton, 1974.

Pic, Paul, *La protection légale des travailleurs et le droit international ouvrier*, Paris: Alcan, 1909.

──────────*Traité élémentaire de législation industrielle. Les lois ouvrières*, Paris: Rousseau, 1909.

Pierrot, Dr.Marc, *Travail et surmenage*, Paris: Temps nouveaux, 1911.

Poperen, Maurice, *Syndicats et luttes ouvrières au pays d'Anjou*, Angers: Laval, 1964.

Poulot, Denis, *Le Sublime*, Paris: Maspéro, 1980.

Proudhon, Pierre-Joseph, *La Pornocratie*, Paris: Lacroix, 1875.

Réal, Claude and Rullière, H., *Le Tabac et les allumettes*, Paris: Doin, 1925.

Ronsin, Francis, *La Grève des ventres: propagande néo-malthusienne et baisse de la natalité française*, Paris: Aubier Montaigne, 1980.

Séverac, J.B., *Le Mouvement syndical*, Paris: Quillet, 1913.

Sicherman, Barbara, *Alice Hamilton, a Life in Letters*, Cambridge: Harvard University Press, 1984.

Simon, Jules, *L'Ouvrière*, Paris: Hachette, 1864.

Sorlin, Pierre, *La Société française, 1840-1914*, Paris: Arthaud, 1969.

Stearns, Peter, *Revolutionary Syndicalism and French Labor: A Cause without Rebels*, New Brunswick: Rutgers University Press, 1971.

Tardieu, Dr. Ambroise Auguste, *Etude historique et médico-légale sur la fabrication et l'emploi des allumettes chimiques*, Paris: Baillière, 1856.

Thomas, Edith, *The Woman Incendiaries*, New York: Braziller, 1966.

Tilly, Louise A. and Scott, Joan W., *Women, Work, and Family*, New York: Holt, Rinehart, and Winston, 1978.

Trempé, Rolande, *Les Mineurs de Carmaux, 1848-1914*, Volume I, Paris: Les éditions ouvrières, 1971.

Villermé, Louis René, *Tableau de l'état physique et moral des ouvriers employés dans les manufactures de coton, de laine, et de soie*, Paris: Union générale d'éditions, 1971.

Weisz, George, *The Organization of Science and Technology in France, 1808-1914*, Cambridge: University Press, 1980.

Zola, Emile, *Fécondité*, Paris: Charpentier, 1915.

Zylberberg-Hocquard, Marie-Hélène, *Feminisme et syndicalisme en France*, Paris: Anthropos, 1978.

ARTICLES

Baud, Lucie, "Mémoires", *Le Mouvement socialiste* (June 1908), 418-425 in *Le Mouvement social*, 105 (1978): 139-146.

Benzacar, Joseph, "L'ouvrière au XXe siècle", *Questions pratiques de législation ouvrière et d'économie sociale*, (1902): 172-176.

Darcy, Henri, "Etat actuel de la question des accidents du travail", *Congrès des accidents du travail*, (1898).

Douaillier, Stéphane and Vermeren, Patrice, "De l'hospice à la manufacture, le travail des enfants au dix-neuvième siècle", *Les révoltes logiques*, 3 (1976): 7-28.

Dupasquier, Dr. Alphonse, "Mémoire relatif aux effets des émanations phosphorées sur les ouvriers employés dans les fabriques de phosphore et les ateliers où l'on prépare les allumettes chimiques", *Annales d'hygiène publique et de médecine légale* (hereafter called *Annales d'hygiène publique*), 36 (October 1846): 342-356.

no author, "Emile Magitot", *Enciclopedia Universal Ilustrada*, Madrid: n.d.

Ewald, Francois, "La formation de la notion d'accident du travail", *Sociologie du travail*, 1 (1981).

Farge, Arlette, "Les maladies des artisans", *Annales*, (September-October 1977): 993-1006.

Guilbert, Madeleine, "La présence des femmes dans les professions et ses incidences sur l'action syndicale avant 1914", *Le Mouvement social*, 63 (April-June 1968): 125-141.

La Berge, Ann, "The Paris Health Council, 1802-1848", *Bulletin of the History of Medicine*, 49, 3 (Fall 1975): 339-353.

Lécuyer, Bernard-Pierre, "Les maladies professionnelles dans les 'Annales d'hygiène publique et de médecine légale' ou une première approche de l'usure au travail", *Le Mouvement social*, 124 (July 1983): 45-70.

Loubère, Leo, "Left-Wing Radicals, Strikes, and the Military, 1880-1903", *French Historical Studies*, 3 (1963-1964): 93-105.

——————————"French Left-Wing Radicals, Their Views on Trade Unionism, 1870-1898", *International Review of Social History*, 7 (1962): 203-230.

Drs. Magitot, Roussel, Bouchardot, Lereboullot, Laborde, and Vallin, "Sur l'assainissement de la fabrication des allumettes", *Bulletin de l'Académie de médecine*, (February 23, March 2, March 9, 1897) in *Revue d'hygiène publique*, (March 1897).

Magitot, Dr. Emile, "La Fabrication des allumettes et les accidents phosphorées", *Revue d'hygiène publique*, (June 1894): 497-501.

——————————"Des industries insalubres", *Revue des Deux Mondes*, CXL (March 1897): 144-168.

Mattei, Bruno, "La normalisation des accidents du travail.

L'intervention du risque professionnel", *Les Temps Modernes* (January 1976): 988-1006.

Paulet, Georges, *Revue du Commerce et de l'Industrie*, (April 16, 1896): 208, 209, 366-369.

Perrot, Michelle, "Quelques éléments de bibliographie sur l'histoire du travail", *Le mouvement social*, 105 (1978): 131.

Pinot, R. "Les institutions sociales dans la grande industrie", *Revue politique et parlementaire*, (February 1924): 17 and (March 1924): 17.

Potonié-Pierre, Eugénie, "Les allumettiers", *La Question sociale*, 3 (June 1895): 203-204 and (July 1895): 284-285.

Reberioux, Madeleine, "Demain: les ouvrières et l'avenir au tournant du siècle", *Revue du Nord*, LXIII, 250 (July-September, 1981): 661-671.

Roussel, Théophile, "Sur les maladies des ouvriers employés dans les fabriques d'allumettes chimiques et sur les mésures hygiéniques et administratives nécessaires pour assainir cette industrie", *Comptes rendus hebdomadaires des séances de l'Académie des sciences*, (1846), 292-295.

Sauvez, Dr. E., "Dr. Emile Magitot", *L'Odontologie*, XVII (May 1897): 289-299.

Satre, Lowell J., "After the Match Girls' Strike: Bryant and May in the 1890s", *Victorian Studies*, 26 (Autumn 1982): 7-31.

Scott, Joan, "Socialist Municipalities Confront the French State", in Merriman, John M. *French Cities in the Nineteenth Century*, London: Hutchinson, 1982: 230-246.

Sedillot, "Observations de nécroses des os de la face et d'affections pulmonaires survenues à des ouvriers employés à la fabrication des allumettes chimiques", *Comptes rendus hebdomadaires des séances de l'Académie des sciences* (1846): 437.

Tardieu, Dr. Ambroise, "Etude hygienique et médico-légale sur la fabrication et l'emploi des allumettes chimiques", *Annales d'hygiène publique*, VI (1856): 14-52,

Tilly, Louise, "Individual Lives and Family Strategies in the French Proletariat", *Journal of Family History*, (1979): 144-162.

——————"Paths of proletarianization: the sex division of labor and women's collective action", *Signs* (1981): 400-417.

——————"Structure de l'emploi, travail des femmes et changement démographique dans deux villes industrielles: Anzin et Roubaix, 1872-1906", *Le mouvement social* 105 (1978): 33-58.

——————"Three Faces of Capitalism: Women and Work in French Cities", in Merriman, John M. *French Cities in the Nineteenth Century*, London: Hutchinson, 1982: 165-192.

Trempé, Rolande, "Lutte des travailleurs et améliorations de la santé", in lmhof, A.-E., *Le Vieillissement*, Lyon: Presses universitaires de Lyon, 1979.

Vallin, Dr. Emile, "L'assainissement de la fabrication des allumettes. Rapport à l'Académie de Médecine le 9 février 1897", *Revue d'hygiène publique*, (March 1897): 97-119.

——————"L'intoxication phosphorée et les allumettes de Paris", *Revue d'hygiène publique*, (January 1897): 184.

——————"Les Nouvelles Allumettes", *Revue d'hygiène publique*, (1898): 5-6.

Weisz, George, "The Politics of Medical Professionalization in France, 1845-1848", *Journal of Social History*, (Fall 1978): 3-30.

Weiner, Dora B., "Public Health under Napoleon: The Conseil de Salubrité de Paris, 1802-1815", *Clio Medica*, 9, 4 (1974): 271-284.

Zylberberg-Hocquard, Marie-Hélène, "Les ouvrières d'Etat (tabacs-allumettes) dans les dernières années du dix-neuvième siècle", *Le mouvement social* 105 (1978): 87-107.

For Product Safety Concerns and Information please contact our EU
representative GPSR@taylorandfrancis.com
Taylor & Francis Verlag GmbH, Kaufingerstraße 24, 80331 München, Germany